AMERICAN AXES

ERRATA

Some of the bracketed material pertaining to the Daniel Simmons axe manufactory of Berne, New York, on page 137 and 139; the Weed, Becker & Co. axe manufactory of Cohoes, New York, on page 140; and the White & Olmstead axe manufactory of Cohoes, New York, on page 140 and 141, was inadvertently credited to the author. This material should have been acknowledged to Kenneth D. and Jane W. Roberts from their work, *Planemakers and Other Edge Tool Enterprises in New York State in the Nineteenth Century*, published by the New York State Historical Association and Early American Industries Association, copyright 1971, by K. D. and J. W. Roberts.

AMERICAN AXES

A SURVEY

OF THEIR DEVELOPMENT

AND THEIR MAKERS

By HENRY J. KAUFFMAN

For

Myra and James Keillor,

both of whom

love old objects made of iron,

particularly axes.

Copyright © 1972 by Henry J. Kauffman
Copyright © 1994 by Olde Springfield Shoppe

International Standard Book Number: 1-883294-12-6
(previously ISBN 0-8289-0138-4)

Published by
Olde Springfield Shoppe
10 West Main Street
Elverson, PA 19520-0171

ACKNOWLEDGMENTS

At the outset it should be stated that the concept of this book was the brain child of James A. Keillor and the Early American Industries Association. It was they who brought the subject to the attention of the author. The help of many additional hands and minds is evident in the survey. I am particularly indebted to Dorothy Briggs for her sympathetic interest in the subject and for the excellent line drawings she has made for the book.

Also to Henry Glassie and *Pioneer America* for permission to use the drawings of corner-timbering. Among others who helped are:

Willis Barshied	Fred Kinsey
Edwin Battison	James Knowles
Lee Donnelly	Jack Loose
Mrs. Dorothy Duncan	Vincent Nolt
John Grabb	David Perch
Vernon Gunnion	Harold Peterson
Ray Hacker	Dale Pogatchnik
Mrs. Virginia Hawley	Kenneth Roberts
John Heisey	Elmer Stahl
Harry Hetzel	Ray Townsend
Carroll Hopf	George D. Tuttle
Charles Hummel	John Waddell
Donald Hutslar	Wallace Wetzel
John Kebabian	Harold York

I am also indebted to several organizations: The Bucks County Historical Society, Carnegie Library, Connecticut Historical Society, Goschenhoppen Historians, Inc., Henry Francis duPont Winterthur Museum, Lancaster County Historical Society, Western Reserve Historical Society, and the York County Historical Society.

H. J. K.

CONTENTS

INTRODUCTION	1
I EARLY EUROPEAN AXES	5
II THE EIGHTEENTH CENTURY	16
III THE NINETEENTH CENTURY	30
IV THE TWENTIETH CENTURY	52
A PORTFOLIO OF COLLECTORS' AXES	57
V CARE OF THE AXE	101
VI USE OF THE AXE	107
VII DIRECTORY OF AMERICAN AXE MAKERS	115
GLOSSARY	143
BIBLIOGRAPHY	148
INDEX	149

INTRODUCTION

If a hierarchy of woodworking tools were established, it is quite probable that the axe would head the list. In the words of the well-known English authority on woodworking tools, W. L. Goodman, in the first chapter of *The History of Woodworking Tools:*

> Any comprehensive study of the history of woodworking tools must of necessity begin with the axe; it was not only the first, but for many years about the only woodworking tool of any kind, and it was still important right up to the end of the Middle Ages. At the present time [1966] it is still used a great deal for its original purpose, the felling and preparation of timber, and it occupies a modest but useful place in the modern carpenter's kit.

In this one succinct paragraph Mr. Goodman states his case for the importance of the axe from the time of its invention in prehistory until the twentieth century. His focus on the tool covers a wide range of technological change, but, unfortunately, the connoisseur of American axes will find little in his survey which is relevant to the American tool. This is an unfortunate omission because it has been in America that the tool has reached its highest functional development. However, it would be rather surprising to find a European author sensitive to the nuances of axe development in America, or to any facet of American life which did not significantly affect his own. (It is evident today that some reciprocity of recognition is occurring, but the millennium in such matters lies in the very distant future.)

The American development of the axe occurred for two principal reasons. First, the Europeans who came here had never before been confronted with such huge virgin timber as stood here, and although they had good tools for hewing it, they did not have a very satisfactory

Focus on the axe

American Axes

one to fell it. Second, the very fact that the Europeans left their homeland for an uncertain future in the wilds of America attested to the fact that emotionally they were pioneers and, as such, they would be likely to resolve many of the problems encountered here.

Part of the problem of focusing attention on the American axe arises from the fact that the earliest ones used here were made in Europe, and certainly the first ones made here were European in character. Thus, in the earliest colonial times a dividing line could not be drawn between the two categories. As a matter of fact, the object was really a European-American axe. Because iron, unlike wood, is similar regardless of the place it was made, the essential substance of an axe does not help to identify its origin. Short of some identifiable maker's mark, the manufacturers of most of our early axes must remain anonymous.

European influence

It seems certain that most of the first axes made in North America

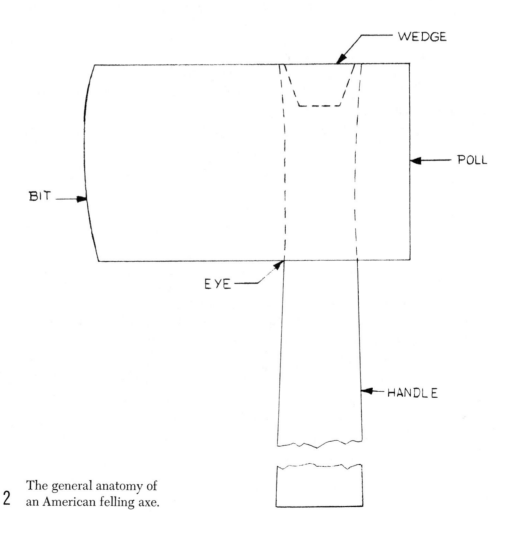

The general anatomy of an American felling axe.

were made and used on the Atlantic seaboard, a few exceptions occurring when trading companies brought in blacksmiths to their centers of exchange to repair and resharpen axes. (It is possible that these blacksmiths also made axes, at least in emergencies.) As settlers moved westward and southward, their needs were supplied by smiths who went with them and were responsive to individual needs. This procedure was the beginning of very high specialization in the forms of axes, a differentiation which was picked up by the big manufacturers in the nineteenth century. The axes were mostly of the felling variety, but there were other purposes for which an axe was needed. As one example, ice had to be cut; and it is quite obvious that the demand for this type of axe was great in New England, but negligible in Georgia. The pace of specialization increased; as evidence of this trend, one manufacturer informed the writer that at one time the company manufactured about three hundred different types. The president of the Mann Edge Tool Company, in Lewistown, Pennsylvania, reported that in 1969 they were producing seventy different patterns; however, the bulk of their production involved only about twenty.

Although a friend of the writer residing in Canada reported recently that he went into the countryside to have a blacksmith make an axe for him, it is not likely that such a procedure could be duplicated many places in the United States. Nationwide, less than a half-dozen manufacturers are making axes today, in a spirit of competition which might be described as "lively." In recent years axes have been imported from Canada, and today sizable quantities are coming from Sweden, home of one of the largest axe manufacturers in the world. Over a million axes are produced annually in America today.

The intent of this survey is not to present a definitive study of axes. Rather it is an attempt to focus attention by means of narration and photography on the history of the tool, with major emphasis on its development in the United States from the colonial era to the twentieth century.

Doubtless some enthusiasts will notice that all the categories of axes made within the designated time span are not included. As has been noted, at one time one manufacturer was producing three hundred different axes. There were only slight differences between many of them, and such refinement could not be detailed in a survey of this size. Thus, I have selected for inclusion some of the common shapes, as well as bizarre forms which only connoisseurs are apt to see. By contrast, the selection should be both interesting and informative.

Introduction

American specialization

American Axes

The depths of metallurgy have not been probed, for each manufacturer had (and has) trade secrets in manufacturing axes. Such information is difficult, if not impossible, to obtain. However, different materials were used to make axes in different eras and comment is made about them in the glossary.

Finally, the reader should know that despite the fact that the axe has been an unnoticed tool in the past, it is now eagerly sought by collectors all over the country. It is not uncommon for an axe to bring several hundred dollars at an auction, and such prices are often surpassed in private transactions. The tariff is raised for exotic forms and those bearing signatures of identified makers. For this reason I have compiled and appended a selected Directory of American Axe-makers.

Collector's item

It is hoped that this presentation will appear worthwhile, and lead others into a deeper and wider study of the subject.

September 1971 HENRY J. KAUFFMAN
Lancaster, Pennsylvania

Early European Axes

Information about axes in all periods of time is fragmentary. Unfortunately, only one researcher has concentrated his efforts on this important tool, and therefore bits of information about it must be gathered from various archaeological reports which are concerned chiefly with other matters. Thus a survey of axes used before the settlement of the Western Hemisphere is necessarily brief, and can be encompassed in a single chapter. Later chapters will treat in more detail the manufacture and use of axes in America.

It is of particular interest to note that the evolution of the axe was not chronologically uniform in all areas. For example, the Stone Age in Mesopotamia, in Western Europe and in the Western Hemisphere did not coincide. When Columbus arrived in America, Europeans had been familiar with iron for about two thousand years, but the American Indian was then engaged in making tools of stone.

It is improbable that the earliest stone tools—called "eoliths"—were purposely fashioned by early man; he apparently found stones shaped by the forces of nature so that they could be used for specific functions. To kill a sleeping animal, a large heavy stone could be held in the hand. One with a sharp edge could be used to scrape fat and hair from the hide. Eolithic tools were not hafted: the human arm provided the power and leverage to perform the desired function. The axe *per se* did not exist in this earliest era.

Eoliths

A number of innovations occurred in the Lower Paleolithic period. Man discovered that rocks could be shaped and made sharp by chipping the edges, a process usually described as flaking. When they became dull, they could be resharpened by removing another layer of chips. This knowledge led to shaping tools for specific purposes, and tools such as drills, axes, and chisels were deliberately shaped for the

American Axes

first time. The craftsmen who fashioned these tools had an uncanny ability to produce beautiful patterns by flaking, as well as very attractive symmetry in the shape of their product.

But possibly the most important invention of the era was the attaching of handles of antler or wood to these improved tools. The first handles for axes were in the shape of a bent knee, the end to which the stone was attached being short, the other being long to provide leverage in using the tool. The center of the antler was sometimes hollowed to receive the stone, bitumen (mineral pitch) was used to "seat" it properly, and it was fastened securely by wrapping thongs around both stone and handle. If the handle was wood, the stone was inserted in a split in the short end of the limb or laid against a flat surface; either way thong again was used to lash the two parts together. By modern standards these tools did not serve very long, but similar ones are used today in a number of undeveloped countries. Stone tools of this type are called "celts."

The first handles

Although a flaked tool was superior to any used before, the practical bent of man led to the discovery that a flaked tool ground smooth was better still. Such tools were made by flaking them to their approximate size and shape, after which they were abraded by rubbing them on another stone. These tools are quite sophisticated in form, many resembling the shape of modern tools.

Possibly the greatest asset of ground or smooth stone tools was the fact that they could be attached to a haft in a number of ways. The old method of inserting the tool in the split end of a knee-shaped stick was doubtless continued, for transitional procedures often involve both old and new methods. It is also known that in some cases one end of the stone was ground to a point and the stone inserted in a hole in the handle. Although this jointure was also strengthened by using bitumen and thongs, it is evident that the handle was likely to split, rendering the axe useless. Another procedure for inserting a haft was to drill a hole through the stone by means of a bow-drill, a device made of wood and thongs, and to which a stone drilling point was attached. The holes were drilled halfway through the stone from each side, usually meeting with a precision which seems amazing today. Sometimes grooves were ground on each side of the stone for a handgrip, or to accommodate the split ends of a limb which was then bound to the stone with thongs. Axes found in America show that this was the method of the American Indian.

Joining the haft

Although progress in the evolution of the axe is measured in cen-

Bronze Age axe found at Bradfield St. George, Suffolk, England. (John S. Kebabian collection)

turies, or possibly millennia, man did finally discover that pieces of native copper could be pounded into useful tools. The malleability of this metal permitted it to be pounded into a variety of shapes, but its softness minimized its utility for tools requiring a sharp edge. This inadequacy was in part overcome by the discovery of bronze, a combination of copper and tin. Bronze is also reasonably malleable, but the ease with which it can be cast into desired shapes made it greatly superior to copper, so that copper was discarded as a substance for making tools.

Using bronze

Once again the new resembled the old. The first bronze tools were very similar in shape to their stone and copper predecessors. Because tools like chisels and axes could not be cast with a very sharp edge, such an edge was produced by pounding the metal. This procedure also made the metal hard and produced a wider cutting edge, such as is found on a modern axe. The bronze bit was still lashed to a bent-knee haft, and when in use it twisted out of line because of the jarring impact with the wood and the inadequacy of the binding arrangement. This was overcome by pounding edges along the sides of the bit and fitting the handle into it. In addition, an offset was cast on the flat side of the bit to keep it from slipping upward on the handle. Thus the haft was enclosed on three sides and the bottom. The final step was to dis-

American Axes

Palstaves

cover the use of a core so that a socket could be cast into the end of the axe and the handle be completely enclosed. But there was a difficulty with this technique. When the handle dried, it became smaller and loose in its socket. The problem was solved by casting an eye near the top of the socket through which thongs were laced to keep the handle in place. These bronze tools were called "palstaves."

Although it is evident that substantial progress was made from the

Bronze shaft-hole axes and adzes from Asia Minor and the Greek Islands from 2900 to 800 B.C. (From *The History of Woodworking Tools* by W. L. Goodman, courtesy of the David McKay Company)

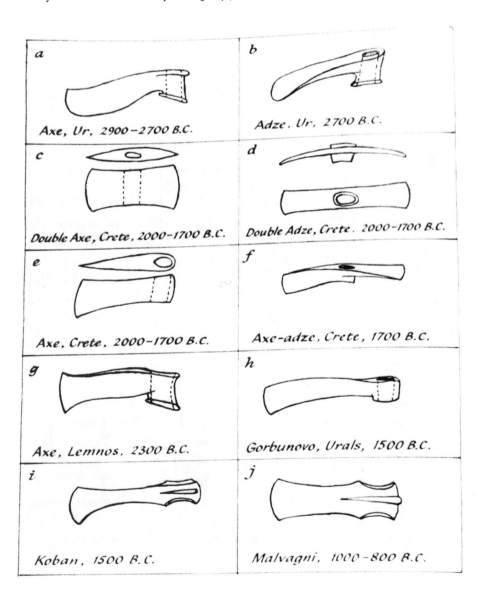

Early European Axes

eoliths to the palstaves, further improvements in the metal and method of hafting were necessary before a satisfactory tool was evolved. Because the technology of processing harder metals was then unknown, early improvements were in the direction of more efficient hafting. The shattering impact of the axe against the end grain of the handles was finally eliminated by mounting the haft, or helve, at a right angle to the major axis of the tool. This procedure was made possible by using the previously learned technique of coring to create an appropriate opening at one end or in the middle of the axe. Both the single- and double-bitted bronze axe were made by this method. The practice spread widely, and examples have been found in Sweden, Crete, China, Hungary, Siberia and France.

The first axes of iron were doubtless replicas of the highly developed forms in bronze. Presumably they were made completely of iron, for it was obviously better suited than bronze to the demands of the tool, although less satisfactory than later examples made of iron with a steel bit. They were probably forged from one piece of metal by a blacksmith; although some examples demanded that forging be supplemented by welding.

The Middle Ages

The artist in the Middle Ages seems to have been eager to portray axes, for their forms and uses are found in woodcuts, illuminations and stained-glass windows, and in the famous Bayeux Tapestry depicting the Norman conquest of England.

Possibly the most important form was the axe with a round or elliptical eye and a long narrow bit; it was widely used for felling either trees or adversaries. Surviving examples indicate that those for battle were inlaid with narrow strips of silver or gold; thus killing became a more aesthetic exercise than chopping down trees. Examples of this type frequently have a poll, or head, as counterbalance, a feature which seems to have disappeared until it showed up again later in the American felling axe of the eighteenth century.

The Middle Ages was also a period of fanciful shapes in axes, one of the most interesting being the shape of a fan. Although this tool is frequently considered a weapon, it is illustrated in manuscripts as a side, or hewing, axe. In the Bayeux Tapestry (*circa* A.D. 1080) it appears to be used for squaring a piece of timber and for smoothing the hull of a ship. The length of the blade suggests that it was cumbersome to use, and at a later date the front lobe was eliminated, thus giving it an asymmetrical shape, which Goodman called a "bearded pattern."

The front edge of the bearded pattern was straight away from the handle with the entire cutting edge extending toward the rear. This

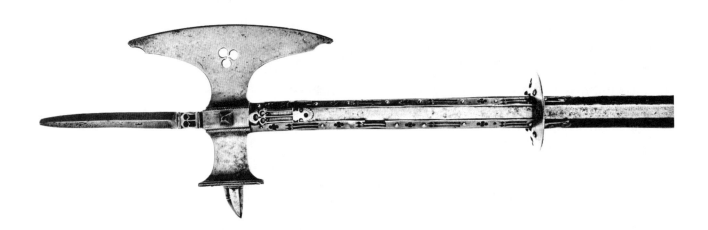

Above: a Gothic pole-axe used as a battle axe, and having the tradition of once belonging to Edward IV of England. Including the shaft, it is 70 inches in length, with a total weight of 6 pounds, 7 ounces. *Below:* an early seventeenth-century heading axe in the Tower of London. Black and rough from the forge, the blade is set at an angle. (This, and the photograph of the pole-axe, courtesy of The Master of Armouries, Her Majesty's Tower of London)

Early European Axes

change added efficiency by bringing the cutting edge under the end of the handle, which was straight and either long or short. It is believed that some of these axes had a canted handle so that the hewer would not scuff his knuckles as he trimmed the side of a log.

Another popular type was a T-shaped tool, made of two pieces of metal. A strap wrapped around the handle formed the eye and extended downward where a second section was welded to the first at a right angle. The one illustrated here had a long cutting blade which appears to have been made completely of iron.

The variety of forms given to axes in the Middle Ages suggests considerable experimentation, with the result that much charm was achieved in form. However, a sharp modern tool was still in the future.

As has been pointed out, information about historic axes is not plentiful. There is comparatively more data available about stone axes than about any other category until modern times. Possibly the period we know least about is that of the transition from European patterns to those indigenous to North America. One might hope that some of the fanciful designs of the Middle Ages could be identified as products of American craftsmen, but the writer's attention has never been called to such a discovery. On the other hand, hundreds and possibly thousands of a small-bitted, poll-less axe, called a "French trade axe," have been found in various parts of the United States. It has been reported, possibly facetiously, that they were so plentiful in some areas of New York State that a farmer placed hooks on his plow handles to carry those he excavated in the normal tilling of his land.

French trade axe

The origin and use of this small axe, which Europeans occasionally called a hatchet, is somewhat obscured. There does not seem to have been any specific use for such an axe in Europe, where there were doubtlessly several better axes for felling and hewing. Perhaps it served as a utility tool in rural areas. To the American Indian it seems to have been the *pièce de résistance*. A great many have been found in Indian graves, tokens of respect, for giving one up must have been a great sacrifice; any tool of iron was of much importance to the Indian community.

Calling the tool a "trade axe" implies that it was made in France and was brought here by French traders to exchange for furs and other commodities available from the Indians. These axes were also known as "Biscayan" because many were made from iron in Northern Spain. It is believed that axes, made there in large quantities, were sold to nearby neighbors, who in turn shipped them to America. The nearness of this locality to the Atlantic seaports of France increases the likeli-

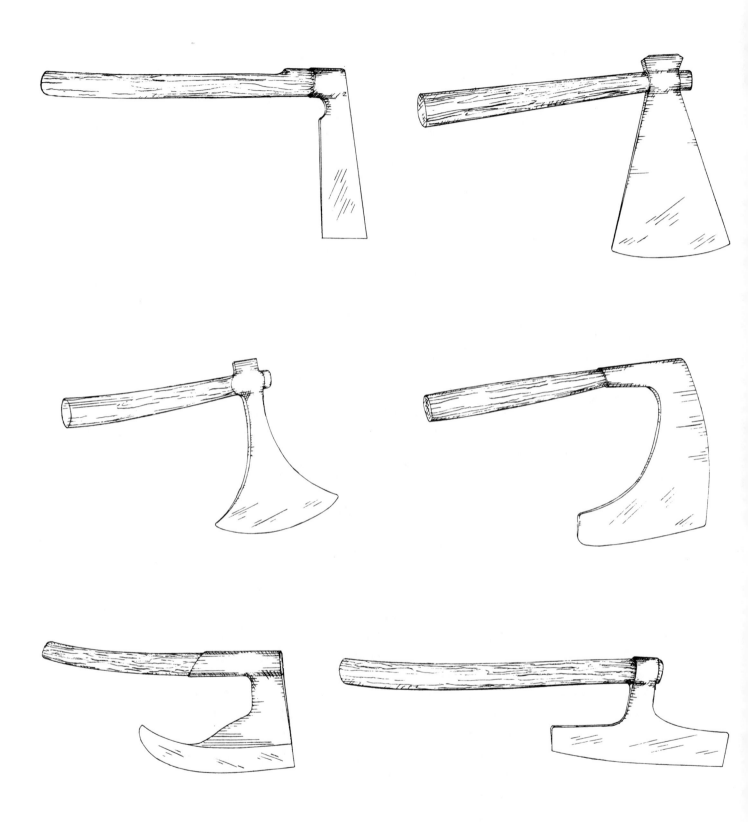

Typical axe shapes of the Middle Ages.

Early European Axes

hood of this occurrence. Little research has been done in regard to these tools, but it is reported by Carl P. Russel in his *Firearms, Tools, and Traps of the Mountain Men* that:

> Hints regarding the distribution of the Biscayan axe by Spaniards in America are found in some reports written by or for commanders of the early sixteenth-century expeditions in the country that now makes up the southeastern part of United States. During the 1520's, Lucas Vasquez Allyon led several forays into the Carolinas and Florida. He made special but futile efforts to colonize the country near the mouth of the Savannah River in 1526. Allyon and many of his followers sickened and died there after a few months residence. Included in the properties that they introduced and abandoned to the local Indians was the Biscayan axe.

It was reported by a member of De Soto's expedition that Biscayan axes were indeed found in the territory in which Allyon met his fate. Later, both French and Spanish explorers brought similar tools to Florida.

Axes in trade

It was in the area around the St. Lawrence River that the French traders became the most successful in the latter portion of the seventeenth century, and the good relations they established with the Indians there were conducive to an extensive interchange of goods. This friendship promoted trade between tribes to the west around the Great Lakes, with the result that many Indians came to own an axe long before any member of the tribe had ever seen a white man.

In the seventeenth century English traders penetrated much of the Hudson Bay area as well as New England and other regions. They were meticulous in their demands for Biscayan axes to trade with the Indians, and the so-called "Hudson Bay axe" bears a kinship to the one from Biscay. English, as well as French and Dutch, axes have been found as far west as Green Bay; some have been found in Indian graves at Washington Boro, Pennsylvania, having been buried long before William Penn took over his claim for the great Commonwealth. The sides of English axes are flat, while those of the French and Dutch have round holes with a strap around the handle forming a big round eye.

In retrospect, the small European trade axe does not appear to have been a very useful tool; however, its importance to the American Indian seems evident. The Stone-Age technology of the Indian certainly limited the size of his undertakings, and an iron or iron-and-steel axe helped him resolve many of his problems. One of the most important

Trade axes of the Susquehannock Indians, found at the Ibaugh excavation in Washington Boro, Pennsylvania. Other artifacts at the site indicate that the burials occurred in the seventeenth century. (Courtesy of the North Museum, Franklin and Marshall College)

uses must have been to cut and trim saplings for his wigwams and bark-covered long houses. He could have used it to shape household objects such as bowls and mortars, and he certainly could have made good use of the iron tool in dispatching a wild animal or an enemy. If the axes were of bright metal, the lustre of the tool must have intrigued him, and one can easily understand why the axe was popular with both young and old. How its shape happened to evolve will probably never be known.

It is likely that the earliest colonists brought axes with them as well as materials from which others could be made here, but the axe of the seventeenth century in America lies beyond the area of hypothesis. The only lead concerning the shape of the axe in the late seventeenth and early eighteenth centuries is an example of a carpenter's axe illustrated in Moxon's *Mechanick Exercises* published in London in 1703. This axe looks like an oversize model of the earlier trade axe, with a straight handle about two feet long. It could well have been the utility axe of the colonists, and probably served until axes better adapted to American conditions were made by American smiths. A reasonably similar example is found in the *Diderot Encyclopedia*, although this publication appeared about a half-century later than Moxon's. Both axes probably had some influence on the course of events in America.

Early European Axes

Colonial utility axe

Carpenter's axe from *Mechanick Exercises* (1703), compared with what could be a utility axe from the *Diderot Encyclopedia* (c. 1762) with the eye squared off to create a rudimentary poll.

II

The Eighteenth Century

In the eighteenth century an American pattern began to emerge in many facets of life, and this change became especially evident in matters relating to the style and manufacture of axes.

The casual observer of today's technological procedures for making axes would be impressed by the simplicity of the operation, but in the eighteenth century it was a formidable exercise. Today the supply of the metals used in axes is virtually unlimited, whereas making an axe then required precious commodities which were scarce in America. Not only were iron and steel limited in quantity and reasonably expensive, but considerable ingenuity and skill were required of the blacksmith who made them.

The blacksmith In the earliest settlements the blacksmith was undoubtedly the most important craftsman. It is known that several were among the first group which came to Jamestown, and they were probably of equal importance in the early settlements in Massachusetts and Pennsylvania. So necessary was this tradesman that, when a town was laid out, a lot near the center was designated as the site for a blacksmith shop and given free to the smith who would build a shop there and carry on the trade for a prescribed number of years. Of course, much time was spent in ways other than making axes or laying a new piece of steel on the edge of an old one, but these were certainly some of the most essential of the tasks he performed.

It is entirely logical that some of the early colonists brought axes with them, and, as pointed out earlier, metal was imported to make additional axes. The British mercantile system, not as restrictive then as later, permitted supplies of iron and steel to be shipped to America. A letter written to his homeland by the Reverend Francis Higginson, the first minister of Salem, Massachusetts, suggested:

Eighteenth Century

> Before you come be careful to be strongly instructed what things are fittest to bring with you for your more comfortable passage at sea, also for your husbandry occasions when you come to land. For when you are parted from England you shall meete neither fayres or markets to buy what you want. Therefore be sure to furnish yourselves with things fitting to be had before you come as : meale for bread, malt for drink, woolen and linen cloth, and leather for shoes, and all manner of carpenters tools, and a great deal of iron and steele to make nails, and locks for houses, and furniture for ploughs and carts, and glass for windows, and many other things which were better for you to think of there than want when here.

The cost of imported iron, as well as the uncertain supply, must have caused many smiths to cast about for local sources of metal. The simplest way to produce iron was in a bloomery resembling a Catalan forge, used in Spain as early as the tenth century. The name arose from the small amount of iron called a "bloom" which was so produced. The bloomery was like a large blacksmith's forge with a deep firepot, the blast being introduced into the side instead of the bottom. Sometimes several small blooms had to be formed to get a usable quantity of metal, but bloomeries could be profitably operated at first because of the small cost and minimum of labor involved in their erection. They were used for working small surface outcroppings of ore, or to explore the resources of a region where the building of a large furnace was contemplated.

The bloomery

The bloom never became hot enough to become a liquid and had to be heated and hammered several times to take out most of the extraneous substances in the ore. In the process the iron was decarbonated and the product made malleable and easy to fabricate into useful objects. It should be noted, however, that all of the impurities were rarely removed. Fibrous striations and scabs remained and are often evident in the finished tools. The presence of these in a tool does not guarantee its production from bloomery iron, however, for later products of forge and triphammer often have similar imperfections.

The growing demand for iron, combined with an advanced knowledge of metallurgy, caused the first operative furnace in colonial Amer-

American Axes

ica to be built near Boston, at Saugus, Massachusetts, in the 1640's. The exterior shape was square and tapered inward toward the top. The interior shape resembled two cones abutting each other at their widest parts, the top one being about twice the height of the lower one. The area where they met is called a bosh. Most furnaces were subsequently built in a similar shape. They were built against the side of a hill so workmen could easily get to the top and dump measured quantities of iron ore, limestone, and charcoal into the furnace. The intense heat within, created by the burning charcoal and fanned by a blast of air from water-powered bellows, fused and melted the contents until the bottom of the furnace was filled with molten iron, on top of which floated a layer of slag.

After the proper quantity of molten iron had collected on the hearth of the furnace, it was tapped at ground level. The molten iron flowed away from the furnace into a big ditch dug in a bed of sand, and was then channeled into smaller ditches. The iron of the big ditch was called a "sow" and that in the smaller ditches "pigs," since the pattern formed by the hot red metal suggested a sow with suckling pigs.

The cooled cast iron was hard and brittle. To make it usable by the blacksmith, it was taken to a nearby facility called a forge, very similar in form and function to the bloomery. In the forge the metal was con-

Charcoal iron verted to a pasty mass as intense heat further purified it, principally by removing the carbon which caused it to be hard and brittle. Finally, blobs of metal were beaten into a bar—in later times it was taken to a rolling mill where the metal was reduced in size—which the blacksmith could handle in his small forge.

This so-called "charcoal iron" was a wonderful medium for the blacksmith. It was malleable and easy to weld, two qualities essential for making axes, and the blacksmith took full advantage of them. Throughout the eighteenth century axes were made mostly of iron with a steel bit to retain a sharp cutting edge which could be resharpened a number of times. When the steel edge was entirely worn away, the axe was taken to a blacksmith who "laid" a new edge of steel on it —a bit of steel that can be easily seen on many axes from this era.

America achieved a certain self-sufficiency in the production of iron in the eighteenth century. (It is pointed out in the *Encyclopedia of American History* that "the Colonies produced 1/70th of the world's iron supply in 1700, but in 1775, 1/7th more than England and Wales combined.") But the process of making steel was more perplexing. Acrelius, a Swedish traveler in America, describes a steel furnace he visited in 1756 called the Vincent Steel Works on French Creek in Chester County, Pennsylvania:

> At Branz's [apparently the operator] works there is a steel furnace, built with a draught hole and called an "air oven." In this, iron bars are set at a distance of an inch apart. Between them are scattered horn, coal, dust, ashes, etc. The iron bars are thus covered with blisters [after lengthy heating], and this is called "Blister Steel." It serves as the best steel to put upon edge tools.

Eighteenth Century

Blister steel

The production of steel must have been a precarious business, for there was certainly no method for precise control of the amount of carbon the iron absorbed. Because the carbon was contained in the surface blisters, a blacksmith had to fold and weld the bars many times before there was a fairly uniform distribution of carbon throughout the steel. The inadequacy in quality or supply of American steel is evident in the fact that on June 2, 1780, John Ott advertised as follows in the *New Jersey Gazette:*

> The subscriber takes this method to inform the public that he has a large quantity of best German steel, and that he intends to apply himself wholly to making axes in the neatest manner which will be warranted. Any person who will apply may depend on being supplied at as cheap a rate as the times will permit.

Although this advertisement suggests axe-making was an established trade in America in the eighteenth century, there is little evidence to support the theory. A sizable number of signed "goosewing" axes survive from that era, but virtually none of the makers have been identified with a certain place or time.

Ledgers of general blacksmiths reveal that most of them did some work connected with the making or repairing of axes. John Miller, a blacksmith working in Lancaster, Pennsylvania, recorded that in the year 1762 he did the following:

> Steeled 9 axes, including one Dutch broadaxe
> Made 50 ditching axes
> Upsett one axe
> Made a post [mortising] axe from an old one
> Made one meadow hoe
> Made two hatchets

After examining hundreds of axes, the writer is convinced that many methods were used to fabricate this important tool. It is evident that

American Axes

the earliest metal axes used in America, the trade axes, ranged in size from those that could be slung on a belt to a larger type used in the forest and commonly called "felling axes." Most of them have a band wrapped around the handle; the heads have no poll, and the bits fan out in length and width according to the method of fabricating them. The fact that the band around the handle was thicker than the bit suggests that this axe was unsuitable for splitting or hewing. Although the sides of some trade axes are flat and straight, the characteristic silhouette form has survived until the twentieth century in what are now generally called "Spanish axes."

Examination of many trade axes has failed to reveal precisely how they were made. However, assumptions can be drawn from a general knowledge of manufacturing methods and the basic procedures for making any axe.

Pattern to weld

A bar of iron slightly less than twice the intended length of the axe was cut from a pattern and the mid-section thinned with the cross peen of the blacksmith's hammer to wrap around the handle. This method was economical in the use of metal, and the loop which was to fit around the handle was relatively easy to form. A slanting scarf, tapered inward, was formed at both ends of the iron pattern so that, when brought together, there would be a V-shaped slot for inserting a piece of steel to make the cutting edge of the bit. The next step was to bend the slab of metal around a mandrel, or an eyepin, to shape the opening for the handle and bring the ends of the bit together for welding. This meant bringing the iron to a glowing, sparkling heat and then quickly welding it on the anvil. (It should be pointed out that it was impractical at this point to have the steel touching any portion of the iron: not only would its presence have made the welding operation cumbersome, but the steel would probably have been decarbonized—and thus softened—destroying the possibility of producing a sharp edge.)

The flux

Fixing the wedge of steel in the iron bit was a very delicate operation, and the flux was important. For welding iron, clean river sand, or powdered sandstone, made a good flux, but they were not satisfactory for welding steel, or steel to iron. For welding the two metals together, a glass-like composition of borax with sal ammoniac added in a ratio of 2 to 1 was used, as was well-dried and finely powdered white potter's clay, moistened with salt water or brine.

The difference in the welding heats of the two metals was also very critical. The small piece of steel was heated to a cherry red, and the iron to a white heat, for joining. Then they were sprinkled with flux

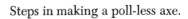
Steps in making a poll-less axe.

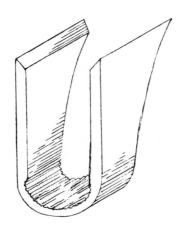
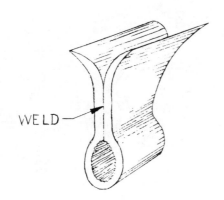
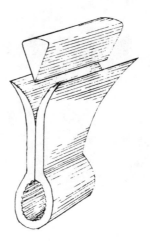
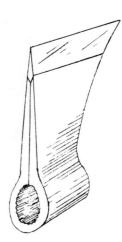

American Axes

and the temperature again increased to a welding heat. By skillful and careful pounding the iron and steel were then forged into one integral piece of metal, forming the bit of the axe. The forging on the end of the bit thinned and widened the edge so that little grinding was needed beyond what was required for shaping.

Because the bit now tapered to a thin edge, knife-blade fashion, it was impossible to create another V when the old wedge of steel was ground away and a new one needed. The blacksmith then "laid" a flat

Edging with steel piece of steel along one side of the iron bit and welded the two together. The steel, being thin, could be ground to a knife-blade edge. The result, however, was a blade that was all steel on one side but had iron on the other side ending within a short distance of the edge.

The craftsmen developing the tool industry in America were trained in traditional European methods of making such articles as guns, plows and axes. However, they soon began developing widely accepted innovations, creating new tools such as rifles and axes which were superior to those being made in Europe. Under the pressure of the new needs in their new land, they made tools which were not only different but *better*.

Anyone who wields an axe sooner or later finds himself using the area opposite the bit as a hammer. This practice is evident in the marks

A transitional shape—late seventeenth or early eighteenth century—between the trade axe and the American felling axe. The embryonic poll seems to have been formed when the axe was made. (Author's collection)

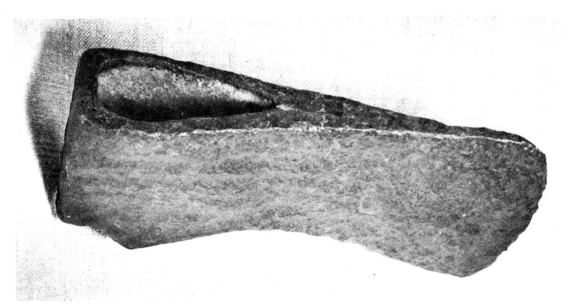

of wear on a large trade axe in the collection of the Pennsylvania Farm Museum of Landis Valley. Since its rounded surface was ill-suited for hammering, some unsung craftsman squared the back of the axe to create a poll. At first the poll was little more than a strap with slightly squared corners, as illustrated by an axe in the author's collection. The forged blank for such an axe might appear similar to that of the trade axe, but the final shape was remarkably different. The sides were long and straight, which improved the tool for splitting and even made it usable for hewing, if one had no other axe. The shape of the eye was changed from an ellipse to a triangle, and its size, in proportion to the size of the axe, considerably enlarged.

These alterations proved so successful that the next step was to enlarge the flat poll, thus—incidentally or deliberately—improving the balance of the axe for use in felling trees. With a more delicate balance between the metal on each side of the handle, the axe became a refined tool almost beyond improvement.

With this enlarged poll came a new mode of fabrication. Two slabs of iron were cut from a piece of stock, each half as thick as the intended axe and having very much the same rectangular shape. A crosswise recess was forged into each piece near one end, allowing a short, flat portion to form, eventually, a sizable poll. Slanting scarfs converging inward were formed on the other end of each piece, so that, when the two slabs were welded together, an undersized hole for inserting the handle and a V-shape for inserting the steel wedge were created. There is abundant proof of this procedure, for many such axes have split apart revealing the way they were fabricated.

This type of axe became known as the "American felling axe." It appears to have evolved over an uncertain period of time. However, an advertisement of an edge-tool maker in the *Pennsylvania Packet and Daily Advertiser* for July 7, 1789, shows that the American felling axe was fully developed at that time. As a matter of fact, the illustration closely resembles the "Kentucky" pattern made by the famous Douglas Axe Company and illustrated in their catalogue of 1863. More remarkable still is the fact that the "Jersey" pattern of the 1970's continues to be very similar to the prime model of 1789.

While axes were being made by the early methods described above, a strip of steel was welded to the poll to minimize the possibility of breakage and increase utility, so the axe could be used as a sledge. An unfinished example in the Pennsylvania Farm Museum of Landis Valley shows that a strip approximately one-half inch thick was added to reinforce the poll.

Eighteenth Century

Growth of the poll

American felling axe

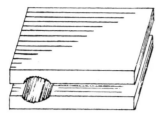

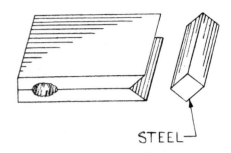

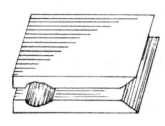

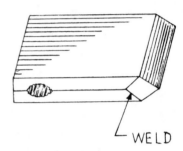

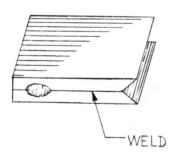

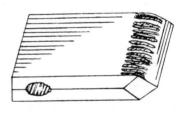

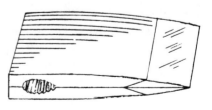

Steps in making the American felling axe.

Eighteenth Century

Wm. Perkins, Blackſmith,
Makes and Sells at his Shop in Water ſtreet, next Door to the Corner, above South ſtreet, in Philadelphia the following ARTICLES; and has now by him a Quantity of the beſt Kind of ———
WOOD or Falling Axes, Broad Axes, Adzes, Carpenters Mauls, Hatchets of different kinds, Ditching or Banking Shovels, Weeding or Corn Hoes, Grubbing Hoes, Tucking Hoes, Chiſſels, Plane Irons, 1od and 12d Nails, Hooks and Hinges, and many other Articles too tedious to mention.   w&f5w

Earliest advertisement found in this survey showing the true form of the American poll axe. (From the *Pennsylvania Packet and Daily Advertiser,* July 7, 1789)

Although most axes made by hand in America were generally fabricated as those described, there were many variations. For example, a number examined for this survey had the usual poll of iron, but a slice of steel was inserted from the cutting edge to the eye. The iron overlapped the steel and the two parts were welded or, in rare cases, riveted together. Presumably, rivets were used when the weld was regarded as inadequate for holding the two different metals together—a doubtful

American Axes

practice since it is likely that constant pounding would eventually loosen the joint.

Thus far, this discussion of axes has centered on two historic types, the trade axe—surrounded with an aura (perhaps merited) of romance from having been found at the earliest historic sites—and the American felling axe, which is justly famed as an outstanding achievement of early American technology. Neither of these axes, however, was well suited for the important task of hewing.

An axe for hewing

Hewing axes were used in Europe quite early; they are sometimes depicted in Medieval illustrations. A German woodcut, which appears in *Ancient Carpenters' Tools* by Henry Mercer, shows a camp follower carrying a goosewing axe; presumably it was used as a military weapon, or in the erection of a temporary camp. Tools of this general shape were brought to Pennsylvania by Germans who migrated there, and a few examples have been found. The writer found one, almost identical to the one illustrated, in central Pennsylvania. The fact that it was with a sizable number of indigenous axes makes its foreign origin suspect, and the clear imprint of a name in typical English letters further supports the theory it was of American manufacture.

Although the shape of the tool illustrated is quite different from that of the common goosewing found in Pennsylvania, they do share some common characteristics. The most obvious is that the handle is canted away from the main axis of the axe. This design protects the hewer, who otherwise would be constantly scuffing his knuckles along the side of the log. It should be noted that the canting of the poll resulted in making left- and right-handed axes. The back side of the axe is flat, and the edge of the opposite side basiled, thus providing a chisel-like action as the axe moves diagonally across the surface of the log. A large number of these axes are signed or have some form of decoration on the side opposite the one which slices along the length of the log. This is a curious circumstance in view of the fact that virtually no felling axes of the eighteenth century bear the name or initials of the maker.

The goosewing axe

The construction of the goosewing is quite interesting. Obviously an apprentice did not begin his first day on the job by forging one. There appear to have been two methods for forming the eye. One was simply to wrap a piece of iron around an eyepin and then to weld the two extending parts together, as was done in making a trade axe. The other method was to rough out two pieces of iron as illustrated and to thin the edges preparatory to welding. The two parts were then placed

Steps in making a goosewing axe.

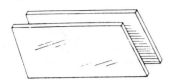

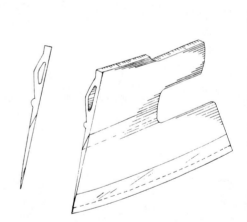

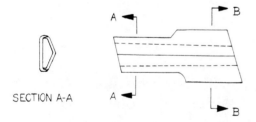

SECTION A-A   SECTION B-B

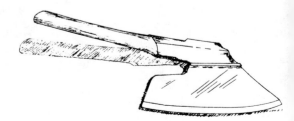

Goosewing axe of the nineteenth century, made of cast steel by G. Sener of Lancaster, Pennsylvania. This is the only goosewing made of cast steel found in this survey. (Courtesy of the Pennsylvania Farm Museum of Landis Valley)

Using an eyepin over an eyepin and welded together in a way similar to the use of a bick iron in welding gun barrels. The eyepin was very necessary: otherwise the eye would have been crushed in the welding process. The eyepin kept the eye open and the taper uniform from the large end to the small end. Most goosewing axes have a ridge parallel to the channel of the eye, which seems to have no function other than that of decoration. After the eye was made, it was welded to a flat slab of iron with the desired thickness and shape of the intended axe. The joining of these two parts is readily discernible on a large number of axes examined for this survey. The shape of the slab determined the shape of the axe, which in turn was dictated by its intended function and the aesthetic tastes of the maker.

Few tools made by American craftsmen show such a variety of interesting forms as the goosewing axe. Some collectors, inspired by their symmetry, call them the "angel wing." Others point out a similarity to halberds of an earlier period. An interesting feature of a few examples is the fact that the maker dated his product, thus adding to its interest and inadvertently aiding the researcher in trying to identify the styles of a particular era.

Eighteenth Century

On all goosewing examples of the eighteenth century a band of steel was welded to the cutting edge. The joint is almost always evident on the back side of the axe, and on some examples it can be plainly seen on the front. In the nineteenth century the goosewing pattern is a rarity; however, at least one was found in this survey. It was marked *Cast Steel*, and was made by G. Sener of Lancaster, Pennsylvania.

To recapitulate, through the eighteenth century axes were made mostly of iron. To make them functional, a band of steel was welded along the cutting edge. This band was as wide as one and a half inches; on a later example a sheet of steel virtually covered one side of the axe, an arrangement similar to that found on plane irons. The plainly visible joints are due to the fact that iron and steel were rarely welded perfectly, and even when they were, the different texture of the two metals defined the joint.

A time of transition

These methods of fabrication were probably carried throughout the eighteenth century and into the nineteenth century, when steel was substituted for iron in making most axes. Numerous examples are now found with the words *Cast Steel* stamped on them, indicating that they were probably made sometime after 1830. These later axes are in demand today, but the earlier goosewing is more interesting and must be regarded as the *chef d'oeuvre* of the axe-maker of the eighteenth century.

III

The Nineteenth Century

All students of technology know that the styles and techniques of the eighteenth century in America were continuations of earlier European practices. The many innovations of the Industrial Revolution caused more axes to be produced, and in most cases they were better. Yet, despite the emphasis on factory methods, axes continued to be made by hand in remote sections of the country in small establishments far into the nineteenth century and, in rare cases, into the twentieth.

Samuel Collins —innovator

A name rarely found in history books among the founders of the Industrial Revolution, and never widely known in America as an innovator, was that of Samuel Collins, who built one of the nation's great industrial complexes of the nineteenth century at Collinsville, Connecticut. Records in the U.S. Patent Office show that he and his employees developed a number of improvements that were obviously a credit to his enterprise as well as a contribution to the economy of the country. His business was not devoted exclusively to the production of axes—the major output being edge-tools and plows—and it survived beyond the middle of the twentieth century. His extensive transactions abroad and the limited development of a market at home might logically explain his obscurity on the industrial scene of America. In 1866, in the evening of his life, he wrote an economic and technological history of his business venture. Excerpts from the first portion are presented here in his own words, to show, among other things, the use of a particular fuel (Lehigh coal) and the specialization of style in utility axes ("Yankee" and "Kentucky").

> *1826* Copartnership formed between S. W. Collins, D. C. Collins and William Wells of Hartford, under the firm name of Collins & Company, for making axes and other edge-tools.

Nineteenth Century

"It has been said that our manufacturing villages have a demoralizing tendency. I wish to show there can be an exception. I would rather not make one cent than to have men go away from here worse than they came."

Sam¹ W. Collins

Samuel W. Collins, founder of a company that operated under his name for more than a century. (Courtesy of Mann Edge Tool Company)

They purchase a sawmill and gristmill and water privilege and a few acres of land on the east bank of the Farmington River in the Town of Canton.

David Collins had a shop at that time in Hartford where he em-

American Axes

The original two-story stone factory of Collins & Company, founded in 1827. (Courtesy of Mann Edge Tool Company)

ployed men in making axes by hand but had no water power.

Not much done at Canton this year, not being able to obtain all the land they wanted on equitable terms.

1827 Commenced quarrying stone for a two-story stone building and for a heavy stone wall west of said building to prevent the river from over-running through that channel.

The forging shop

Built a new forging shop for charcoal fires and a building for storing charcoal. Built three dwelling houses on Front Street facing west, south of Collins Company office.

1828 All our grinding, polishing, blacking and boxing done in the stone building.

Built the first triphammer shop, an undershot float wheel to each hammer, located close to stone building on the east (now a grind shop).

Commenced drawing our patterns this year and making broadaxes with triphammers; had previously made axes from thin ⅝" Russia Iron, sometimes putting a slug in the head as well as the bit. Each man tempered his own axes, forging and tempering eight axes per day, light Yankee. Had not yet made any Kentucky or heavy heads.

1829 Built first shop for Lehigh [coal] fires.

Built a large cylinder bellows standing on and over the waterwheel, the plunger and rod being attached to a crophead moving up and down in slides at each end and driven by long pitmans attached to cranks on the gudgeon of the waterwheel. The air carried underground through middle of the shop under the forge in hollow chestnut logs. No triphammers in this shop.

The men took patterns plated under triphammers (iron and steel) welded and hammered off into good shape for tempering. Ten axes a day's work for foreman and striker, tempered before grinding.

Commenced making Kentucky axes with heavy heads.

Removed the gristmill this year and built a grinding shop on that spot (it being the same now occupied, 1866, by the upper polishing shop, with room for handling machetes in the upper story). Also built two forging shops above last-named shop (north where blacking room is); one for forging axes, the other for repair shop. The wheel for forcing air being in the race under grind shop and the air conveyed in chestnut logs underground.

B. T. Wingate commenced work this year. A very good workman who worked a year making broadaxes by hand without losing a single day, and was considered overseer of the forging and tempering for many years, until he died in 1858.

Altered our bell hours from twelve to ten hours, and gave out some piece work. We had previously worked evenings in winter months. We found the men did just as much work in a day and burnt less coal.

Did not venture this year to make any Yankee axes with Lehigh coal but used charcoal both for forging and tempering.

1830 Forged some Yankee axes this year in Lehigh fires and putting a mark on them, packed a very few in each case, but did not discover that any greater portion failed than of axes made exclusively with charcoal.

1831 Having taken unbound care to make a superior quality of work without reference to cost, the demand increased rapidly. The greatest obstacle to a large increase of production was a want of houses to accommodate and board our men.

Nineteenth Century

The grinding shop

American Axes

Raised the price of our axes to $20.00 per dozen, ten per cent off to the trade, which did not check the demand at all but quickened competition and raised up competitors to bid for our workmen. It is a disputed question whether we lost or gained eventually by advance of price, but I believe it is generally thought to have been a bad move.

1832 In May of this year John F. Wells became a partner in the firm of Collins & Co. taking the place of his brother who died in 1831.

The hardware trade in the cities manifested a strong disposition to push other axes against ours, we sent out several travelling agents this year to show samples of our work and circulate our handbills to the Southern and Western states. By a liberal discount from regular rates we made a sale of axes to Sampson & Tisdale of New York, amounting to $30,000, which is the largest sale we ever made at one time to any firm.

[Certain] men wanted to make arrangements with us to take our whole production, showing their confidence in our business, but it was never our policy to make any "entangling alliances. . . ."

Contracted with David Hinman to build machines for shaping and welding axe polls.

The factory system

Data on axe-making in the early nineteenth century remains fragmentary, but by mid-century some precise data is available on the subject. Meanwhile, one might logically assume that the factory techniques in the first half of the century were probably mechanized versions of methods used by earlier handcraftsmen. These mechanized procedures influenced the industry in a number of ways. They unquestionably increased the quantity of axes made to help meet the rapidly increasing needs of the migration westward, where more and more axes were needed to fell trees and build log cabins. The factory system also contributed to creating a more humane technological environment for the workers. (For instance, the constant pounding of a triphammer must have been sweet music to the men who previously had pounded out axes by hand.) And with the introduction of mild steel as a substitute for charcoal iron, the quality of the axe was also improved. As a matter of fact, the triphammer operators became so clever that evidences of welding cannot be seen on some products. Finally, the shift toward factory production also led to job specialization whereby a workman performed only one of the major operations, such as forging or tempering, instead of producing an entire axe.

Nineteenth Century

The best data the writer has about axe-making in the middle of the nineteenth century—when some axes were still made of both iron and steel, or possibly mild steel for the poll and high carbon steel for the bit—appears in the October 29, 1859, issue of *Scientific American*.

Cutting patterns

After the axe-maker or manufacturer had procured his axe-bars, the first operation was to cut the stock into "patterns," a pattern being the proper portion of metal to make an axe. In making the patterns, a number of bars were kept hot in the forge from which the foreman selected one and placed it on the edge of the hardy on his anvil, whereupon his helper struck it several sharp blows with a cold chisel to score it. After the bar had been well scored by the hardy and the chisel, it was held on the edge of the anvil and the sections were broken away. (Later patterns were cut with a huge shears, which functioned in the manner of hand shears used for cutting cloth.) The patterns were then laid away until they were needed to make axes.

The atmosphere of the forge shop of an axe factory was described by the magazine thus:

> The forge shop has usually the solid ground for a floor, and when everything is in full blast, it affords a tolerable idea of the infernal regions. To a stranger, the roaring flames, the half-naked men, stirring every muscle and perspiring in torrents, the dark recesses of the space now lit up by a sudden glare and now relapsing into their original gloom, the sparks and streams of fire flying angrily in every direction, the horrid and infernal din, the clangor of tools, and the great hammers falling with a tireless, thundering energy, present together a spectacle that seems hardly earthly. No one could easily forget his first experience of such a scene.

Plating-out

To be more precise, this is what was happening in that terrestrial inferno. The first process was called "plating-out." The foreman marked a space about an inch wide across the middle of the pattern; this became what was known as the head, or poll, of the axe. Then two or three patterns were thrown into the forge to be heated to the proper temperature for forging. Next, the pattern was taken to the triphammer to be given its first shape. After another heat it was returned to the triphammer and given a shape which defined the poll.

Finally, the portion forming the eye of the axe was thinned and made ready to be wrapped around a mandrel or eyepin, thus forming a U. The gap between the two cheeks of the U form was filled by placing a throatpiece, a slug or dutchman between the extending members and the parts were welded into one solid mass of metal. The small opening

American Axes

left at the end of the dutchman was enlarged to accept a bit of high-carbon steel. The steel was lightly fastened in place with a few blows, after which the seams or cracks between the two metals were filled with borax and the whole unit returned to the fire. Great caution was needed to prevent complete decarbonizing of the bit by overheating. After one heat at welding temperature, and two or three lower ones, the entire mass was hammered into the finished shape of the axe.

The triphammer

Mr. Collins's history mentioned the use of a triphammer by his company as early as 1828, and it should be noted that this machine was virtually the backbone of the axe industry for about one hundred years. The hammer was mounted on one end of a heavy oak beam, which in turn was mounted on a fulcrum and given an up-and-down motion by cams attached to the axle of a waterwheel. The hammerman placed the red-hot axe on an anvil under the hammer, and by cleverly moving the tool around, a complete and perfect weld could be made. After a final examination the foreman straightened (often pronounced "strightened") the axe and it was ready to be tempered.

The procedure for tempering (in contemporary production grinding precedes tempering) was to heat the whole axe to a red heat and then harden it by dropping the entire tool into water. This process made the metal brittle and, unless otherwise treated, the edge was likely to break in a reasonably short time. To forestall this contingency it was necessary to "draw the temper," thus obtaining a tool capable of withstanding the strain of many impacts and with a hard edge that stayed sharp a long time. Heat was applied to the body of the axe, and by scratching the surface with the end of a file, an array of colors could be

Tempering by color control

seen moving toward the bit. The colors produced were, in succession: yellow, brown, purple, light blue, and black. The shades indicated the desired color for the edges of certain articles: *yellow* for lancets, razors, penknives, cold chisels and miners' tools; *brown* for scissors, chisels, axes, carpenters' tools and pocket knives; *purple* for table knives, saws, swords, gun locks, drill bits, and bore bits for iron and other metals; *blue* for springs, smallswords, etc. *Black* indicated that the metal had been completely annealed (softened thoroughly so as to be less brittle). When the desired color was reached, the entire tool was again dipped into water and the temper fixed.

From the tempering room the axes were taken to the grinding room to finish the surfaces on grindstones that were six to twelve inches thick, four to eight feet in diameter, and weighed from 2,000 to 4,000 pounds.

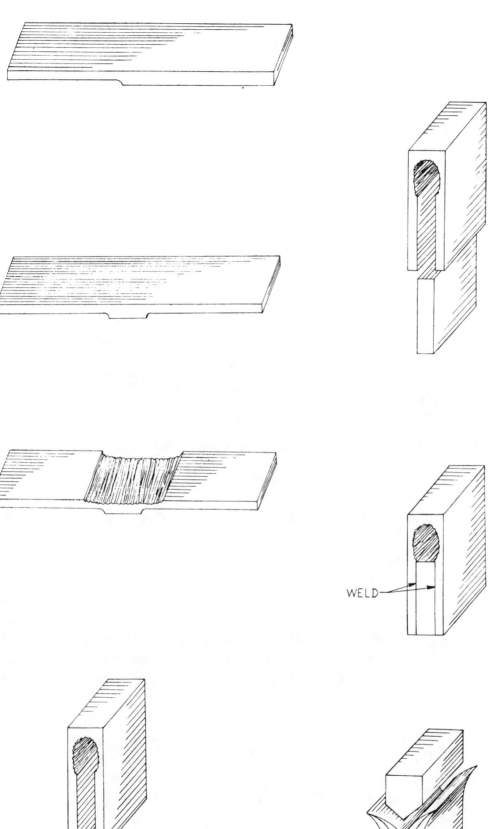

Steps in the manufacture of a utility axe in the mid-1800's.

American Axes

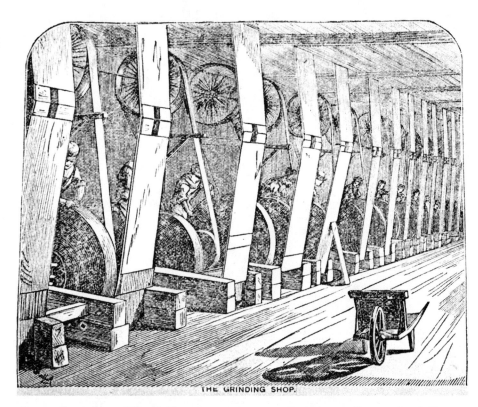

The grinding shop of the Collins Company around 1870. Smaller artificial abrasive wheels in modern times have taken the place of the large natural grindstones shown here. (From *The Great Industries of the United States*)

The grinder seated himself on a "horse" or lever, the back end of which was attached to the floor, with free movement allowed on the front end. The operator placed the axe between the lever and the grinding wheel. By pressing the weight of his body against the lever and skillful management of the eyepin which he held in his hands, he was able to grind the axe quickly to the desired finish. It was necessary to have water drip constantly on the wheel to prevent overheating the axe and spoiling its temper. Since this moisture combined with the dust from the wheel to create an unhealthy condition in the grinding room, it is reported that "no grinders died of old age."

Stamping the mark After that, the soft portion of the axe was stamped with a die to imprint data about the axe or the maker, and the axe was given a high polish on flat wheels covered with leather, on which different grades of abrasive powder had been glued. The finest finish required as many as three or four graded wheels. Sometimes the axes (excluding the sharpened parts of the bits) were painted black or were blackened, as is noted in the 1828 entry of the Collins excerpts. Another source points out that the black finish was prepared by boiling asphaltum in turpentine, which, when applied, dried quickly. These practices were doubtless forerunners of the use of a black finish used by modern manufacturers, one of whom calls the finish "Black Diamond."

Nineteenth Century

Finishes have always been a matter of much importance to axe manufacturers, and they all vied with each other to produce one which was unique and identified their product. By the end of the nineteenth century many different finishes were used. However, one patented by the Mann Edge Tool Company seems to have been a great success, as described in the company's promotional booklet, *Axes*, published in 1897:

Finishing

> This finish was patented on June 2, 1896, and exclusive rights are vested in us. It consists in an exhibition of the temper colors of respective metals in uniformity on the surface thereof wherever exposed, whereby the poll section of the axe presents a lustrous straw or copper color, and the cast steel section, or edge, of the axe presents a lustrous blue color, both colors having been reproduced simultaneously through the action of heat. This finish is valuable in affording an extremely attractive appearance, and in exhibiting conspicuously the structural formation and quality of the axe or tool. For the purpose of distinguishing our XX grade of axes more fully than could have been done otherwise, we have applied this finish to that grade only and shall continue to do so. This finish has proven so popular that our largest competitors have seen fit to offer an imitation, applied to their common goods and offered at a low price, for the manifest purpose of depriving us of advantage. We have, of course, instituted legal proceedings for relief and damages, which are now pending, and we shall protect our rights fully against all imitators.

Any survey of axe-makers and axe-making in America in the nineteenth century is not complete without a brief look at the history of the Mann family and their enterprises. Although several generations of different Mann families were involved in the manufacture of axes, the descendants of William Mann were certainly the most important. The following information is taken from the official *History of the William Mann Family* (in manuscript).

The Mann family —a history

> *William Mann, Sr.*, was born at Braintree, Massachusetts, in 1779. He was an excellent mechanic and manufactured scythes, hoes, forks, cleavers and axes. He died at Axe Mann, Pennsylvania, in 1860.

TO OUR WORKMEN:

After a trial of over two months of KNIGHTS OF LABOR rule we have decided that it does not suit us, and to make a long story short *will never again recognize any Committee of said Organization.* While saying this we would also say we *will always give any one in our employ a respectful hearing.* We have no desire to change wages or time and manner of payment, but in the future will employ and discharge any one we think is not doing his duty. Or, in other words, *will allow no interference with our business from any quarter whatever.* Any one dissatisfied with this arrangement will receive balance due them, and our good wishes for their future prosperity, by calling at the office.

YOURS TRULY,

WILLIAM MANN, JR. & CO.

MARCH 8, 1882.

P. S.—Any one using Threatening Language, Force or attempting to prevent others from working will be dealt with to the full extent of the law.

Declaration of open-shop policy in 1882. (Courtesy of Mann Edge Tool Company)

Nineteenth Century

William Mann, Jr., was born at Johnstown, New York, in 1802. He left Johnstown and located at Boiling Springs, Pennsylvania, in 1824. There he engaged in the manufacture of axes and, one year later, was joined by his brother Harvey, forming the manufactory of William and Harvey Mann.

Small beginnings

In 1833, the partnership was dissolved and, in 1834, William Mann built a small axe factory at Mauch Chunk, Pennsylvania. For power, this plant was dependent on the water from small mountain streams, as were many others of that era. After a very severe winter, William Mann moved to Jack's Mountain, four miles north of Lewistown, and later known as Mann's Narrows. There, he had to start in a small way as he first had to construct a dam across Kishacoquillas Creek. That accomplished, he fixed a raceway and put up a forge with one grindstone, a polishing room and tempering department—all under one roof—where he and his helper made about 10 axes a day.

Customers came to the factory to buy single- and double-bitted axes, until finally his brother Harris went on the road selling axes in various surrounding counties. The factory's products bore a stamp imprinted "William Mann, manufactured near Lewistown, Pennsylvania." His production increased from 10 axes a day in 1835 to 500 a day in 1855. William died in 1855 and was succeeded by William 2nd, J. Fearon, and James, who finally became the sole owner.

"Red Warrior" brand

In 1866, the company under *James Mann* patented the celebrated "Red Warrior" brand of axes and was bought out by the American Axe and Tool Company, then known as the "Axe Trust," for whom James Mann became treasurer. The American Axe and Tool Company was composed of about 16 axe, hatchet and scythe plants, which made fully 75 percent of these tools used in the world. In 1900, the company decided to consolidate their 16 factories; James Mann resigned and re-embarked in the axe business at Lower Mann, Pennsylvania. He died in 1904 and his sons inherited the business under the name of James H. Mann Company.

Since 1904 the Mann Axe Companies have had checquered careers. In 1966 the Mann Edge Tool Company acquired the domestic manufacturing rights of the Collins Company, located for many years at Collinsville, Connecticut, and now operates under its own title as well as that of the Collins Company. Today it is a thriving business producing about a half-million axes per year with the main plant located at Lewistown.

American Axes

Chronologically, the next information found about axe-making appears in *The Great Industries of America*, published in 1872, in which the Collins Company is described as being "the largest establishment in the world for manufacturing axes and edge tools." However, a twentieth-century survey indicates that the W. C. Kelly Axe Company at Charleston, West Virginia, was reputed to have been the largest at that time.

It is interesting to note in *The Great Industries of America* that the Collins Company was reported as still making axes of iron *and* steel in 1872. However, progress from earlier methods is evident:

Mechanization

The heated bar is inserted in an aperture in the machine, whereupon a gigantic knife snips it off at the required length; next a pair of dies give the iron the proper fold or bend; the workman withdraws the lump of iron, inserts it in another aperture, and the hole for the handle is punched; another movement, and it is bent in the opposite direction, and so, by rapid and successive compressions, the head is shaped and ready to receive the bit. This bit, hammered from steel, and finally punched by a die into a shape as long as the axe is wide, with a broad flange left on either side, is now ready to be joined to the iron poll, and complete the form of the axe. The steel is inserted in the iron poll, both being properly heated; the forger turns the two flanges of the poll upon the bit, then runs to a triphammer, under which, by alternate heating and hammering, the two parts are so firmly welded together as to be practically one. When sufficiently drawn out under the triphammer, the next process is to reduce the thickness by grinding; this labor, however, which is slow, expensive and unhealthy for the workman, has been greatly lessened by the introduction of machines which now actually *shave* down the bit of the axe nearly to an edge.

Thus it is evident that the art of axe-making was gradually becoming mechanized. The procedures of hardening and tempering were also speeded up, for instead of heating one axe at a time, a hundred axes were placed in an oven on the periphery of a circular drum, with the bits projecting outward over the edge. The drum was slowly rotated as heat was applied to the bits; the bits of the axes were then cooled and made very hard. They were next taken to a drawing furnace, where workmen controlled the tempering procedure by watching the colors created in the metal by another application of heat. When a pigeon-

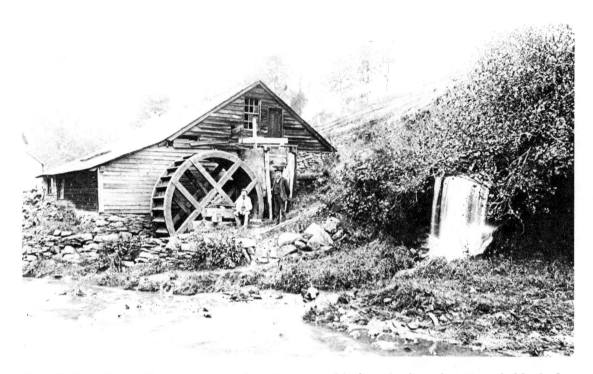

Not all the nineteenth-century manufactories were giants. The Hewes Tool Company—located on Puddle Duck Creek in Lancaster County, Pennsylvania—seems to have been only an over- size blacksmith shop, but it probably had a triphammer operated by the water wheel. (Courtesy of Lancaster County Historical Society)

blue reached the bit, the entire tool was quickly cooled and the temper fixed.

Acute vainglory is evident in the advertising of all makers in regard to their products, from the eighteenth century onward. Each maker pointed out that attention was given to the testing of his axes, and each asserted that his method for testing was best. Simply stated, a typical test was to break a completed axe to see that the proper hardness had been attained; in addition, one properly tempered had to bend a bit before its edge was broken on the edge of an anvil. The following passage from the 1863 catalogue of the Douglas Axe Company is evidence of their concern about the quality of their product:

Testing

> If an axe prove too soft, so as to bend on the edge, or break, in consequence of a flaw in the *Steel* and is returned to the person vending the same within *Thirty Days* from the date of purchase, a new one will be given in exchange. Where the *Steel* is sound, it will be considered as done by harsh usage, and will not be exchanged.

Boston, March 28th 1836

Mr Levi Bailey—

Bought of **JAMES FULLERTON & CO.**
NO. 1 SOUTH MARKET ST. CORNER OF COMMERCIAL ST.
IMPORTERS, MANUFACTURERS AGENTS, & DEALERS IN

Cut Nails.	Pig Iron.	Swedes Steel.	Trace Chains.	Potash Kettles.
Wrought Nails.	American Iron.	Fire Safes.	Sad Irons.	Caldrons.
Tacks and Brads.	English Iron.	Anchors.	Window Glass.	Scythes.
Spike Rods.	Swedes Iron.	Chain Cables.	Shovels and Spades.	English Hoops
Nail Rods.	Russia Iron.	Sheathing Copper.	Cast Steel Hoes.	Hollow Ware.
Nail Plates.	Sheet Iron.	Sheathing Paper.	Common Hoes.	Muskets.
Shoe Shapes.	Cast Steel.	Underhill's Axes.	Anvils and Vises.	Mill Saws.
Sleigh Shoes.	German Steel.	Collins's Axes.	Oven Doors.	Crow Bars, &c.

2	Bars Old Sable	166 lb	5½	$9.13
1	Bar Cast Steel	15	19	2.85
1	German Steel Best	12	14	1.68
1	Round nail Rods	50#	6½	3.25
1	Sho Shapes	50	6	3.60
1	2 inch Nail Plate	50	6	3.00
2	Sheets Eng. Iron	30	8	2.40
2	Bar ½ Round	19	6½	1.24
2	Bars Swed Iron	86	5	4.30
2	put Best Blist Steel	12	15	1.80
9	Bar ¼ + ⅜ Eng	211 lb	4½	9.49
2	Bars ¾ Eng	117	4½	5.26
1	1⅛			

47.40
3 per Disct — 1.42
$45.98

Recd Payment —

Jos Fullerton & Co
pr E. W. Bray

Note the second entry—"1 Bar Cast Steel"—in this invoice of 1836. (Author's collection)

Nineteenth Century

The frequent mention of steel in this excerpt from their catalogue focuses attention on the fact that the Douglas Company was no longer making axes of iron, but rather was using a substance known as "cast steel," which was invented in England in the eighteenth century and probably used in America as early as 1830 for the making of many various tools, axes included. An advertisement of G. Sener in a Lancaster, Pennsylvania, newspaper in 1841 confirms the fact that he was using it at that time. It was also sold in 1836 by a Boston merchant named Fullerton, whose listing also included "Old Sable," a brand of iron imported from Russia, and "the best grade available from any source."

From blister to cast steel

It is known that the lack of homogeneity in blister steel was its major liability, and that this disadvantage could be partially overcome by folding and welding it on a small forge—obviously a makeshift solution to the problem. The answer lay in breaking bars of blister steel to be melted together in a carefully manufactured crucible with an airtight cover. The steel to be melted was carefully selected for uniformity and freedom from undesirable iron cores: all had to come from the same batch, so that the pieces would respond evenly to the heat applied to the crucible. At first, great caution was exercised, and only one crucible was handled at a time.

When the metal reached a fluid state, the contents of the crucible had to be stirred periodically to effect a perfect mixing. Sometimes a number of heatings were preferred to a single lengthy operation; in all cases great care was necessary to avoid the loss of an undue amount of carbon. If such a contingency occurred, the metal reverted to its original iron composition, and therefore could not be hardened and tempered. After the molten metal had been stirred several times in the crucible to ensure its homogeneity, it was poured into a metal mold and formed into an ingot. The ingot was then heated to a low red heat and taken to a triphammer to be forged into a bar. The hammerman placed the ingot on an anvil under the hammer and cleverly moved it around until a bar of solid cast steel was formed.

The list of Fullerton of Boston, mentioned above, includes an item listed as "1 Bar of Cast Steel." It is known that one of the standard commodities of the nineteenth century was an axe-bar about three inches wide, half the thickness of an axe, and about twelve feet long. This product was sold to a merchant or a blacksmith who was familiar with the qualities of the metal and could make usable objects of it. More skill and experience were required to make an axe of cast steel than of the charcoal iron used in the eighteenth century.

Evocative names for patterns linger in these present-day axes advertised by Emerson & Stevens Company of Maine. (Courtesy of Harold York)

The first axes made of steel (mild steel for the poll and high-carbon steel for the bit) were made in a manner similar to those made of iron and steel. The procedure was probably accelerated by the use of swages in the triphammers so that a pattern could be more speedily prepared for shaping and welding. The pattern was then heated and wrapped around an eyepin to form the eye, after which it was reheated to a welding heat and the two protruding parts of the U were welded together. A piece of high-carbon steel was inserted for the bit. This work was done so meticulously that frequently there is no line of demarcation between the two metals involved, but the texture of the two metals is different and the place where they are joined can usually be seen if the axe is cleaned and polished.

Inserting the bit

Much progress was made in the United States in the mass production of axes during the nineteenth century. Possibly the most interesting development of the 1800's—beyond the use of a new metal, steel, and new

Nineteenth Century

modes of manufacturing—was the evolution of a wide range of patterns for felling axes. The Douglas Axe Catalogue of 1863 lists the following: "Kentucky," "Ohio," "Yankee" (it will be remembered that Mr. Collins was making Kentucky and Yankee types in 1830), "Maine," "Michigan" or "Muley," "Jersey," "Georgia," "North Carolina," "Turpentine," "Spanish," "Double-bitted," "Fire Engine," and "Boy's-handled." Later the varieties increased until by the end of the century the list seems endless. Many patterns were made in different qualities to improve the image of the producer (and, incidentally, to confuse the purchaser).

Felling axes —a wide range

Although the professional woodcutter of the 1800's was familiar with the variety of patterns and qualities available, the average collector or observer today is probably best acquainted with the broadaxe of the nineteenth century because examples are numerous, many are marked by their maker, and some have interesting shapes. This tool lacks the charming shapes of the hewing axes of the eighteenth century, but it was probably equally useful for hewing round timber into square or rectangular forms.

A noteworthy feature is that, unlike the hewing axes of the eighteenth century that were made in right- and left-handed models, the broadaxe of the nineteenth century could be used either way. This ambidextrous feature was achieved by the fact that the handle was canted instead of the axe. The versatility of the nineteenth-century tool arose from the fact that it could be adapted for use by inserting the canted handle in either end of the eye. Most of the specimens are basiled on only one side, although examples are found which are ground like a knife instead of a chisel. The latter type was used for shipwork and for forming ties for railroads. And there is always the possibility that an owner, after he no longer hewed with his axe, made a splitting tool of it. A profusion of mortising types and many for special work, such as turpentine and firemen's axes, were made in great quantities by most manufacturers.

The broadaxe

The broadaxe was made in a manner similar to the method for the felling axe, with a few exceptions. Oversize patterns were forged to the desired shape and then wrapped around an eyepin. The parts were brought to a welding heat, and then welded together on a triphammer. The extra size of the patterns provided a large mass of metal which was forged lengthwise on a triphammer until the desired width was obtained. It was then ground to the desired contour. By the end of the century broadaxes were available in patterns such as "Wisconsin," "Western," "Pittsburgh," "Pennsylvania" and "New Orleans."

American Axes

The common small broadaxe usually employed for coopering was probably made the same way. This tool was the adaptation of a larger one, and was designed for a trade which needed only a small tool for use with one hand. These rarely have unusual forms, and are a quite common commodity on the axe market today.

Finally, it might be noted that at the end of the century axes of different grades were available: as witness the following passage from the Mann Edge Tool Company booklet of 1897, cited earlier. Although it describes the axes of one producer, it can be assumed that most manufacturers probably followed a similar procedure.

Varying grades

We manufacture chopping axes of every American or English pattern, both Double and Single Bit of the highest quality and finish. Our "Mann Special" and "Mann's Diamond" brands are made with great care from the best standard materials and are offered as unexcelled in quality and finish by any standard brand on the market. They are sold at the lowest prices consistent with the maintenance of their superior quality and guaranteed to be first class, and free from unusual irregularity, but they are not warranted, each and each, for reasons explained under the subject of "Should Axes be Warranted," contained in this booklet.

There is an expensive method of making steel from stock of an extra high grade which assures a greater degree of strength and uniformity than can reasonably be expected in the standard article. The high cost of this steel makes it impractical to use on axes which are sold in competition with the best standard brands on the market. To meet a demand for something better than the best standard axes we are using this special high grade steel in our Mann's XX Warranted chopping axes of any desired pattern. Axes of this brand are warranted, each and each, to be free from flaws and of correct temper and quality for the service for which they are intended. The steel is guaranteed of a special high grade, costing very much more than the steel used on any standard brand of axes. Axes of this grade, failing through any fault of ours, will be replaced or money paid for them will be refunded, including the freight charges from our customer to ourselves, but not including any other expenses. Purchasers of these goods should make returns at the end of each season, rendering bill covering the same.

Our regular second quality axes are stamped "Lewis Axe Co." and labeled "The Jack Frost Axe." They are furnished in gold

AXE HANDLES

OVAL BENT SINGLE BIT AXE HANDLE—28" to 36"

OVAL SCROLL END SINGLE BIT AXE HANDLE—36"
Used in Mid-West

OVAL STRAIGHT SINGLE BIT AXE HANDLE—36"
Used in Southeast; 36" Lumberman's Pattern.
Standard Straight Axe 30", 32", 34" and 36"
MAINE STRAIGHT 30"-32"

BOY'S SINGLE BIT AXE HANDLE—28"

MINER'S AXE HANDLE—18"-28"

KNOB END FRENCH FAVORITE S. B. AXE HANDLE
28"-30"-32"

OVAL DOUBLE BIT AXE HANDLE—30" to 36"

ADIRONDACK DOUBLE BIT AXE HANDLE—30"-32"

Grip Sizes
STANDARD ....1-7/16" x 13/16"
SLIM ....1-3/16" x 3/4"

ADZE AND BROAD AXE HANDLES

HOUSE CARPENTER'S ADZE HANDLE—34"
ADZE EYE HAZEL HOE HANDLE
(36" Pattern No. 3053   42" Pattern No. 3065)

RAILROAD ADZE HANDLE—34"

SHIP CARPENTER'S ADZE HANDLE—34"

WRECKER'S ADZE HANDLE—36"

OCTAGON BROAD AXE HANDLE—36"

REVERSIBLE BROAD AXE HANDLE—34"

DELTA BROAD AXE HANDLE (Knob End)—34"

LEFT OR RIGHT HAND BROAD AXE HANDLE—36"

Hickory is still the best wood for making axe handles, and is used in a wide variety of shapes. (Courtesy of Hartwell Brothers, Memphis)

bronze finish only. Goods of this brand are not made to order, and therefore are sold only subject to our supply or production. They are slightly defective axes found in our regular product; not seriously imperfect, but hardly equal to our first quality standard. In finish and material they are equal to first quality brands, and are cheap at the price they are sold. They are guaranteed to be high standard second quality goods, but they are not warranted.

Guarantees

In addition to improved materials and techniques for the making of

American Axes

axes, another facet of the industry must be noted for its unique development in the nineteenth century: specifically, the mechanized methods for making handles.

At the outset it should be noted that hickory always seems to have been the preferred wood for handles. In the eighteenth century the wood was probably split to obtain a straight and continuous grain, a necessary quality in a tool handle. Surviving handles from the 1700's disclose that most of them were straight; however, the short handle on some goosewing examples did have a swell near the end of the handle. This feature was doubtless an attempt to improve the grip on the handle, but a straight handle seems to have been common for most of the eighteenth century. There was probably no specialization then in making handles, and most axe-makers no doubt made their own.

Making the handles

The general mechanization of the Industrial Revolution and the preference for curved handles in the nineteenth century must have prompted men to try to devise a machine for making axe handles. This step was almost necessary, for the contoured handle was difficult to make by methods used in earlier times. It is evident that the machine eventually used for turning axe handles was an adapted form of the contour lathe invented by Blanchard earlier in the nineteenth century for turning gunstocks. One such machine was manufactured by Hoyt and Brother Co. in 1885, and was advertised thus:

> The Ober, or Buckeye Handle, Lathe [is] a most wonderful Machine when seen in operation. It is, as its name indicates, a special handle lathe, and yet its range of work is not confined to handles alone. In lengths it turns from end to end 3½ feet, and is equally rapid and perfect on spokes, whiffletrees, etc. or any odd and peculiar pattern, as it is on axe, adze, pick, hammer, and hatchet handles, etc. It will turn at the same time two articles of entirely different kind, as, for instance, an axe helve and pick handle, or a spoke or a whiffletree, and the entire work is so smoothly and well done that it is easily finished on an ordinary sand belt.
>
> The lathe is capable of doing an immense amount of work. The fact that it occupies less room than the ordinary Blanchard Lathe, or the Bailey Gauge Lathe, and that it turns two pieces at the same time, or twice the quantity of any other Lathe of its class, is greatly in its favor.
>
> The Lathe Head may carry two, three, four or six Knives (four preferable) and is so constructed that the Knives cannot cut or eat

into the timber beyond a given depth. This feature (patented) alone has great value.

Size of Pulley on Cutter Bead, 4½ x 5, and should run 3,600 revolutions.

Weight, 1,800 pounds.

In 1887 the H. B. Smith Machine Company of Smithville, New Jersey, was manufacturing another machine which they called a "facsimile" lathe. This one performed essentially the same operation as the "Buckeye," but their literature points out that while one handle was being cut, a second piece of wood could be inserted without stopping the lathe. This feature doubtless increased the utility of such a machine (although the relative merits of the two are not being assessed in this survey). There were possibly still others that successfully performed the cutting of irregular shapes for various tools, guns, shoe lasts, decoys, etc.

Nineteenth Century

Speeding the task

IV

The Twentieth Century

It might be thought that the twentieth century would be the least interesting of all periods in which axes have been made in America if one assumes that in this century axe-making has become a highly mechanized industry. But such an assumption, as will be shown later, would be incorrect.

Because mild steel resembles iron in color and texture, and because a triphammer functions in a manner very similar to a hand hammer, many axes of the nineteenth century appear to be products of an earlier era; many are described by "experts" as hand-forged when, in fact, they were forged on a machine. It is virtually impossible to draw a precise line between products made completely by hand-methods and those forged on a triphammer. As a matter of fact, the use of triphammers continued into the middle of the twentieth century, and one must conclude that patterns and techniques are frequently not as old as they appear to be. (The writer saw triphammers in use in Pennsylvania in the 1950's, and they were probably used in Maine in the 1960's.)

Old and new methods

It is often difficult to understand why old methods linger when new ones are available, but this practice in axe-making has a logical explanation. The cost of a drop forge had to be reckoned against the cost of a number of triphammers. Since the axe business in recent years has been comparatively small, with reasonably small capital involved, the high cost of a drop forge was delayed as long as possible. Besides, there has been a strong conservative tradition of holding on to old techniques —if old, *ergo* they are good—particularly, as long as they stay competitive with the new. Thus the triphammer survived.

General view of the Collins Company plant at Collinsville, Connecticut, in 1926. (Courtesy of Mann Edge Tool Company)

Thousands of axes have continued to be made in the twentieth century by wrapping a piece of mild steel around an eyepin, welding the members together, and adding a bit of high-carbon steel. There was at least one major innovation, however: the throatpiece for the bit was replaced by the "overcoat." There seems to have been a lively debate on the merits of overlaid steel versus inserted steel bits, with the crux of the matter revolving around the size of the bit rather than the method. The following excerpt is from a late nineteenth-century catalogue of Robert Mann & Sons:

Overcoating

> The advantage in favor of the overlaid steel is a larger steel surface and a stronger connection between the steel and the iron.
>
> The advantage claimed in favor of the inserted steel is ⅛ in. more available steel. Practically, however, this is of no importance, as in our experience of 40 years we have never seen an axe of either inserted or overlaid steel, where the steel had been used up to the iron.
>
> We explain the two methods on account of unjust comparisons which have been made of Light overlaid and Heavy inserted steel, and because of misrepresentations being frequently made of overlaid steel not having a sufficient depth of solid steel for wearing purposes, and while this may be true, and doubtless so where a light steel is used in the manufacture, as in the comparison, It Is Not so in regard to our goods, as we use the same weight

American
Axes

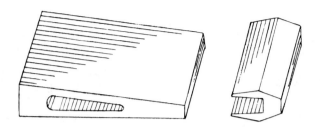

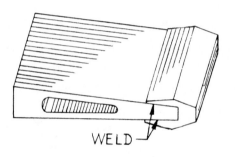

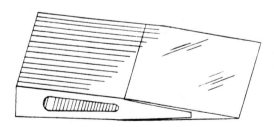

54 Steps in the twentieth-century "overcoat" method of making a polled utility axe.

of steel on overlaid axes as on inserted. The fact is we have several customers who have used our overlaid steel axes for 20 years, and they could not be induced to change to inserted.

Twentieth Century

We make our comparisons from regular size overlaid steel and regular size inserted steel, just as we make goods and leave it to our customers to decide which method is preferable. We have no preference as far as manufacturing is concerned, as the cost to us is exactly the same.

At least for some time the buyer had a choice, but eventually the overcoat method became the most widely practiced.

The fact that the two-piece axe stayed competitive as long as it did gives some support to the claim that this axe was a good one. The literature explaining the production of these axes emphasizes that, since they receive the attention of several highly trained craftsmen, it naturally follows that they must be good.

The battle of the hand- versus the machine-made axe seems to be decided, however, and today axes are made with gigantic drop forges. Billets of steel, each long enough to make three axes, are heated in a furnace to a forging temperature of 2,350 degrees Fahrenheit and manually fed into the drop forge, where, in "the wink of the eye," they are transformed into the shape of an axe. Excess metal, or flashing, that occurs where the two dies meet, is removed by another less powerful drophammer. The eye is then punched with one swoop in a machine called an upsetter, and the tool is ready for the first grinding which roughs the profile of the axe into the desired shape. Next, an eyepin is placed in the eye and the entire surface of the axe ground on a "horse" or "pony"—a vestige of the nineteenth-century practice that continues, probably because it is still the best way to grind an axe.

The drop forge

That part of the bit to be tempered receives a finer polish on a high-speed abrasive belt and is then heat-treated. The first step is to heat the bits in a modern furnace vat of red-hot lead to a temperature of 1,490 degrees Fahrenheit. They are then quenched in a salt brine to make them very hard. The final step of the heat treatment—a delicate operation—requires heating in another lead bath to a temperature of 450 degrees Fahrenheit. After this the bits are properly cooled to draw the temper and then given a final polish by a fine abrasive belt. The axes are then painted by clipping, spraying, or brushing, the handles are inserted, and they are ready for shipping and selling.

Despite these innovations, it is doubtful that axe-making will ever

American Axes become completely automated. To the writer, after making a study of the men who made them and of their methods, the question seems almost irreverent. Moreover, the use of chain saws, which seems to be increasing, is limiting the size and importance of the axe business.

The old tools have charm, they were skillfully made, and they invoke rich memories. Furthermore, it is probable that no other tool has ever played so important a role in the development of nations as the axe has in North America.

Modern pattern for a double-bitted axe, found in the burned-down remains of the Peavy Company of Oakland, Maine. Artifacts of this type are very rare. (John S. Kebabian collection)

A Portfolio of Collectors' Axes

NOTE ON THE PORTFOLIO

The following American axes—fifty-two examples from sixteen collections—attest to the uniqueness of certain indigenous forms as well as to the versatility of a tool which reached the peak of its development as our successive frontiers moved Westward. Some are products of noted manufactories; more are the handiwork of individual axe-makers, most of whom are anonymous craftsmen who created variations to fill the special needs of customers who visited their blacksmith shops.

For convenience, these axes are grouped in seven broad categories: trade axes, goosewings, broadaxes, hewing axes, felling and utility axes, shop axes—comparatively small hand tools used for particular types of woodworking jobs—and, finally, special-function non-woodworking axes. In a number of instances the forms and functions overlap, as will be seen in their descriptions.

This portfolio is presented with the hope that readers of this survey may be prompted to examine axes in their possession or to discover and preserve examples that might otherwise be lost, and thereby add to further knowledge of this important field of Americana.

Trade axe in the style of the seventeenth century. Its battered poll could indicate that it was the precursor of the polled axe of a later period. (Courtesy of the Pennsylvania Farm Museum of Landis Valley)

The example at the top is a trade axe excavated in New York State and bearing the lobed mark found on a number of trade axes. Below is a modern all-steel reproduction of a seventeenth- or eighteenth-century trade axe whose edges show the remains of flashing, suggesting that it was formed by a drophammer. The word "Forge" can be distinguished as part of the maker's mark. (Top, the Willis Barshied, Jr., collection; bottom, the author's collection)

Of unknown origin and function, this axe has an overlaid steel bit that could mean it was made in the eighteenth century. It might be from a much earlier period, however, with the steel merely a replacement. (Courtesy of Colonial Williamsburg, Department of Collections)

The name "H. H. Stricker," poorly positioned and upside down, could be that of the maker of this goosewing axe—although sometimes the owners had their names stamped on their tools. (Courtesy of the Pennsylvania Farm Museum of Landis Valley)

The decorative edge of the goosewing at the top is quite rare. Probably made in Pennsylvania in the eighteenth century, with its undecipherable maker's name struck twice with a die. Below it is an unusually attractive goosewing with its original handle; seventeenth or early eighteenth century. The maker's name is too blurred to read. (Top, courtesy of the Mercer Museum; bottom the author's collection)

GOOSEWING AXES 63

Two more of the many variations found in the goosewing. The lower example is about 9 inches long, doubtless made in the eighteenth century for a special function not known. (Top, the Elmer Stahl collection; bottom, courtesy of the Mercer Museum)

Broadaxe probably made and used in Pennsylvania. Dated axes are very uncommon, and the "1835" on this one is incorporated into the decorative design. Such patterns were made with very simple tools. (Courtesy of the Pennsylvania Farm Museum of Landis Valley)

BROADAXES

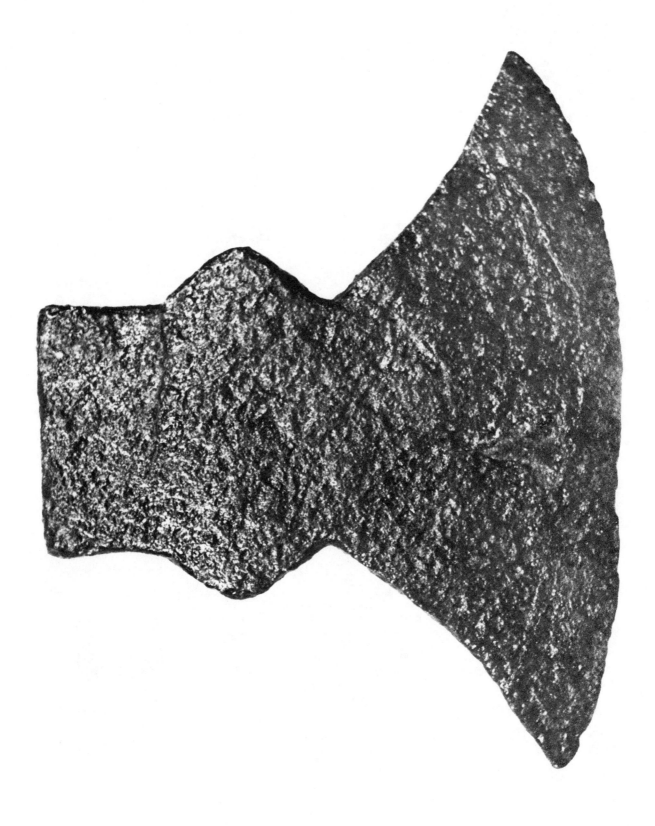

This broadaxe is a vestigial form from the Middle Ages. Made of iron and steel and double-beveled, it possibly had a short handle for use with one hand. (Dale Pogatchnik collection)

Broadaxe of the eighteenth century marked with the initials "I.B." and having a 12-inch cutting edge and a handle 20 inches long. The ridge below the handle is typical for broadaxes of this period, and is a decorative—rather than functional—detail. (Author's collection)

BROADAXES

The steel bit of this broadaxe was made from a file. Although files were often reshaped and adapted as knives because of the high quality of their steel, this is the only example of such use in an axe that the author has seen. (Kenneth Roberts collection)

A broadaxe in the "New Orleans" pattern made by Dunlop and Madeira, whose edge-tool works in Chambersburg, Pennsylvania, is first recorded in 1837. (Author's collection)

At left, an early nineteenth-century broadaxe. At right, a broadaxe made by G. Sener of Lancaster, Pennsylvania, about 1850. (Left, the William Bowers collection; right, the Vincent Nolt collection)

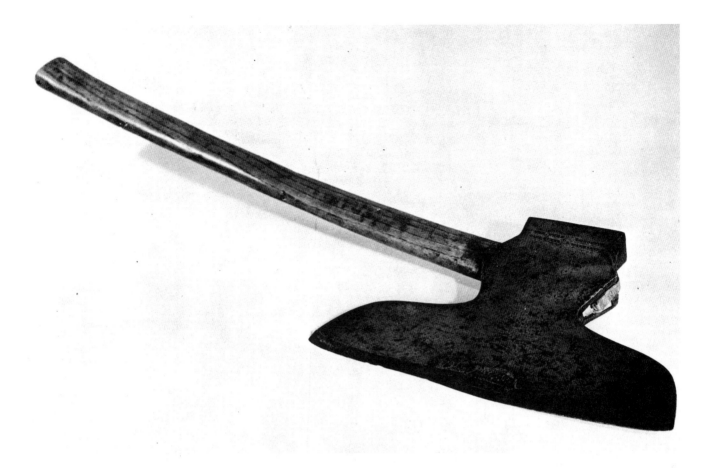

The upper broadaxe is an excellent example of the craftsmanship of the axe-maker and is marked "Wm. Weed, Cohoes, N.Y., Cast Steel." Below it is a rare shape for a broadaxe. (Top, the Kenneth Roberts collection; bottom, courtesy of the Mercer Museum)

BROADAXES

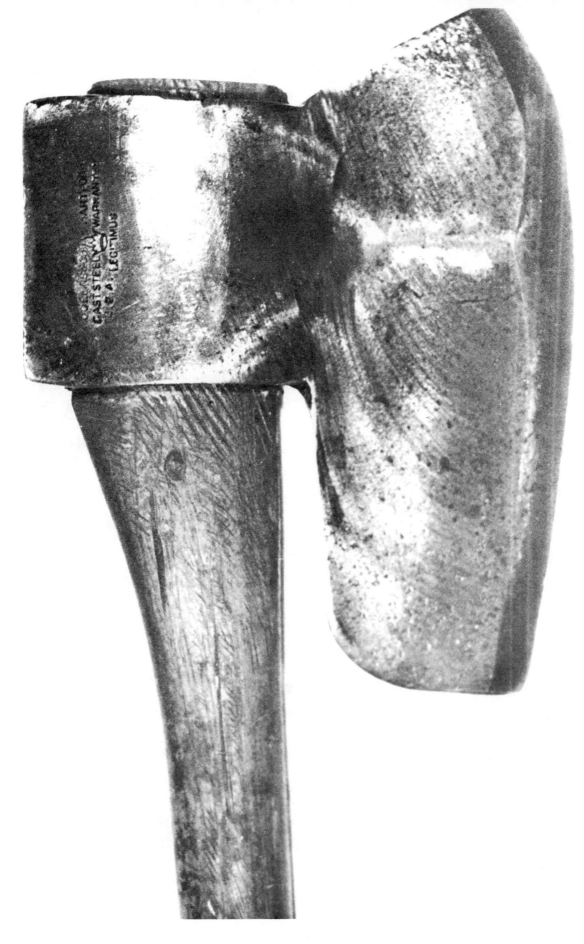

The form of this broadaxe from the Collins factory, Hartford, Connecticut, appears to be much earlier than 1875, when the Collins trademark first was used. (Dale Pogatchnik collection)

Two hewing axes. The upper one, of the late nineteenth century, is marked "Beatty, Chester [Pennsylvania]" and has near the poll the outline of a cow, a symbol this maker used on some of his products. The single-beveled example below is made of iron with a steel bit, and is a highly unusual form. (Top, the author's collection; bottom, the Dale Pogatchnik collection)

HEWING AXES

The rib and hollows are major qualities of this "Perfect" Kelly axe and improved its function for felling. Stamped on its reverse side is a lengthy advertising spiel *(see Page 127)*. (Author's collection)

74   FELLING AND UTILITY AXES

This "Black Raven" is also a Kelly felling axe, from the works of the American Fork & Hoe Co., Charleston, West Virginia. The countersunk holes are not part of the design, but were made later for an unknown purpose. (James A. Keillor collection)

FELLING AND UTILITY AXES

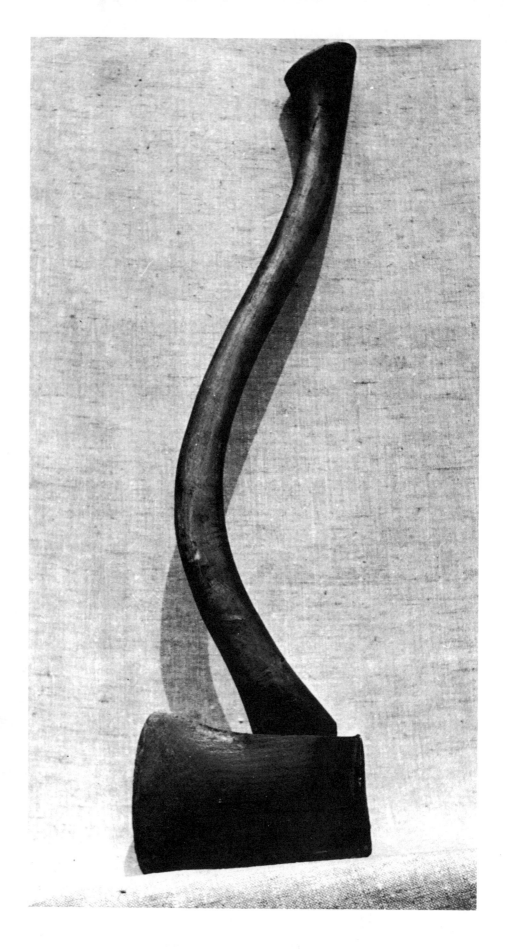

The unusually pronounced curve in the helve of this common wood-chopping axe testifies to the individuality of the man who "hung" it. (James A. Keillor collection)

A rare type of axe—probably eighteenth-century Pennsylvania—used primarily for splitting wood, and which Mercer called a "Holzaxe." The poll is 2½ inches wide, with the total length of the axe 7 inches. The protruding stub at the end of the poll has been found in similar axes, but its function is not known. (Author's collection)

At top, a felling axe with a shortened handle and stamped "Douglas Axe Mfg. Co., Cast Steel, Warranted, Mfd. by W. Hunt." In their catalogue the Douglas Company says some of their products were stamped "W[arren] Hunt," but offers no further explanation. Below is a round-polled felling axe, possibly of the nineteenth century, whose bit has been ground away by sharpening. (Top, the John S. Kebabian collection; lower, the Dale Pogatchnik collection)

FELLING AND UTILITY AXES

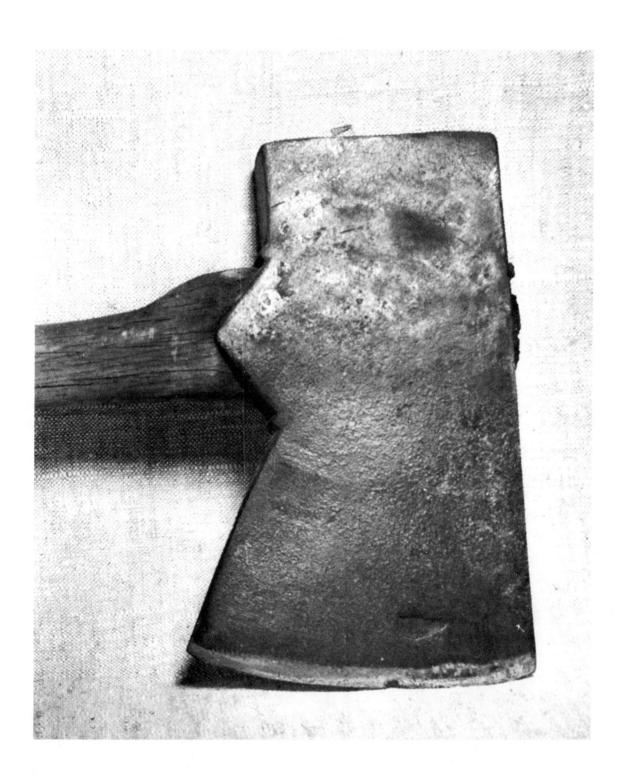

A "wedge" pattern felling axe—but the unmarked poll shows no sign of having been struck. (Courtesy of the Pennsylvania Farm Museum of Landis Valley)

FELLING AND UTILITY AXES

This most unusual heavy hatchet, made by the Douglas Axe Company, has one side overlaid by two thin pieces of steel, with the poll covered with another steel strip a quarter-inch thick. (Author's collection)

The markings on this hand axe are important: few axes are found bearing the names of Ohio craftsmen, and this is stamped "H. Knapp, Cincinnati"; and it is the earliest known dated tool—1831—marked "cast steel." (Raymond M. Smith collection)

FELLING AND UTILITY AXES 81

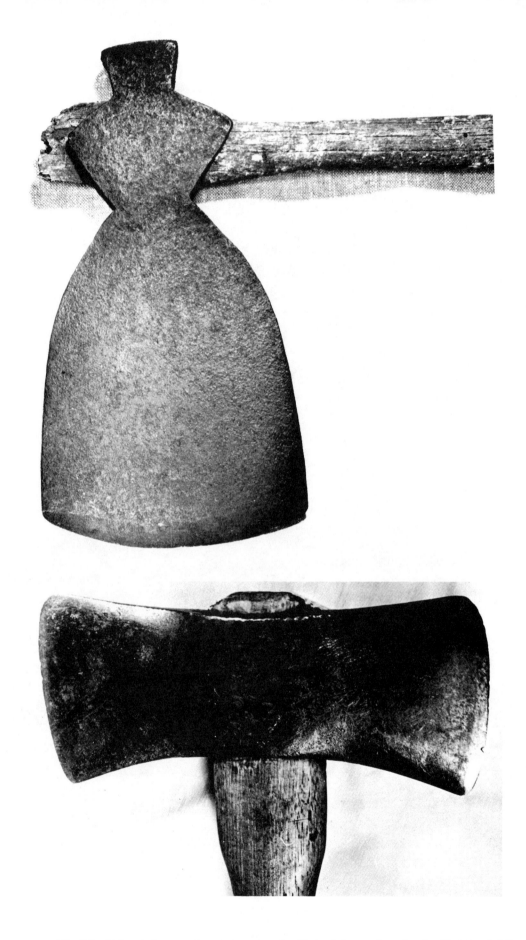

The "English type" at top is quite rare in the United States; but the double-bitted felling axe was common, although very few made before the twentieth century have survived. (Upper, courtesy of the Mercer Museum; bottom, the James Knowles collection)

82   FELLING AND UTILITY AXES

The axe at top, made by Beatty in the late nineteenth century, is the standard cooper's axe. Below is a rare combination: a hewing shape belied by a bulging eye and double bevel. (Top, courtesy of the Mercer Museum; lower, the Dale Pogatchnik collection)

SHOP AXES

G. Sener made this mortising, or posthole, axe in the mid-eighteenth century. (Courtesy of the Pennsylvania Farm Museum of Landis Valley)

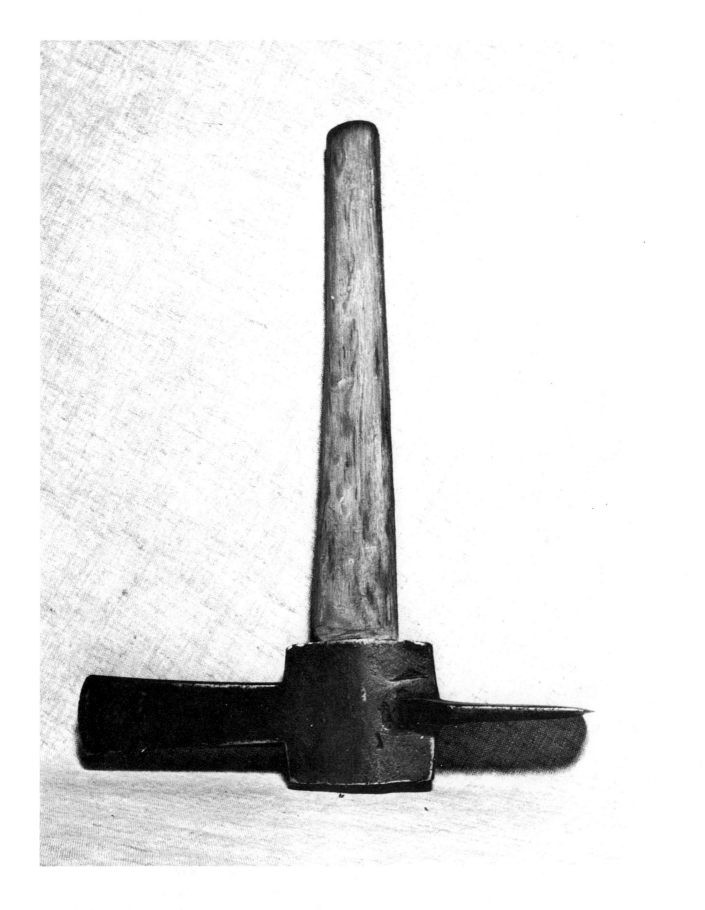

A rare type of mortising axe, c. 1850, is marked "Brady, Lancaster, Pa.," and has its edges at a right angle to each other. (Vincent Nolt collection)

SHOP AXES

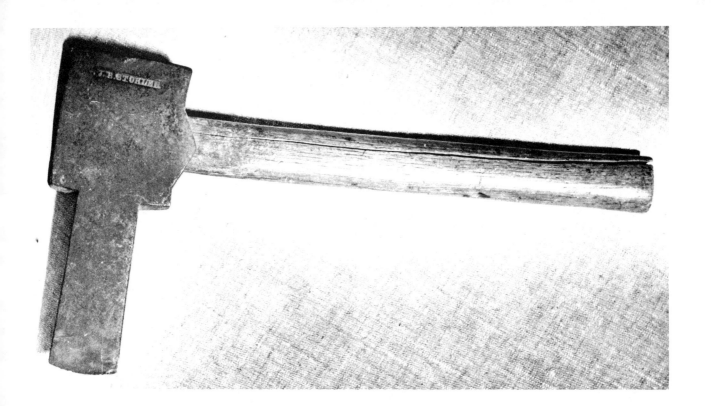

Another Pennsylvanian—J. B. Stohler of Schaefferstown—probably made the posthole axe at top from an ordinary felling tool. Below is a mortising axe without a squared poll, possibly eighteenth century and therefore quite uncommon. (Top, the author's collection; lower, courtesy of the Mercer Museum)

Doubtless a mast-maker's axe, and reduced by frequent grinding—although it is still about 15 inches long. (Vincent Nolt collection)

The 6½-inch handle is but one unusual feature of this shop broadaxe of the 1820's: it is sharpened only on the lower edge, but is beveled all round. (John S. Kebabian collection)

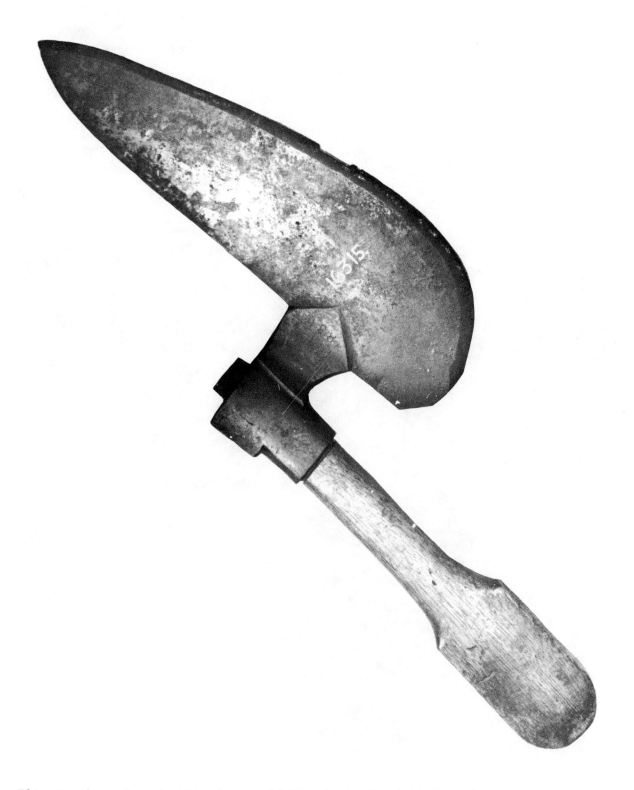

This extremely rare form of small hand axe, used for shop hewing, is probably Germanic and the inspiration for many goosewings made in Pennsylvania. (Courtesy of the Mercer Museum)

SHOP AXES

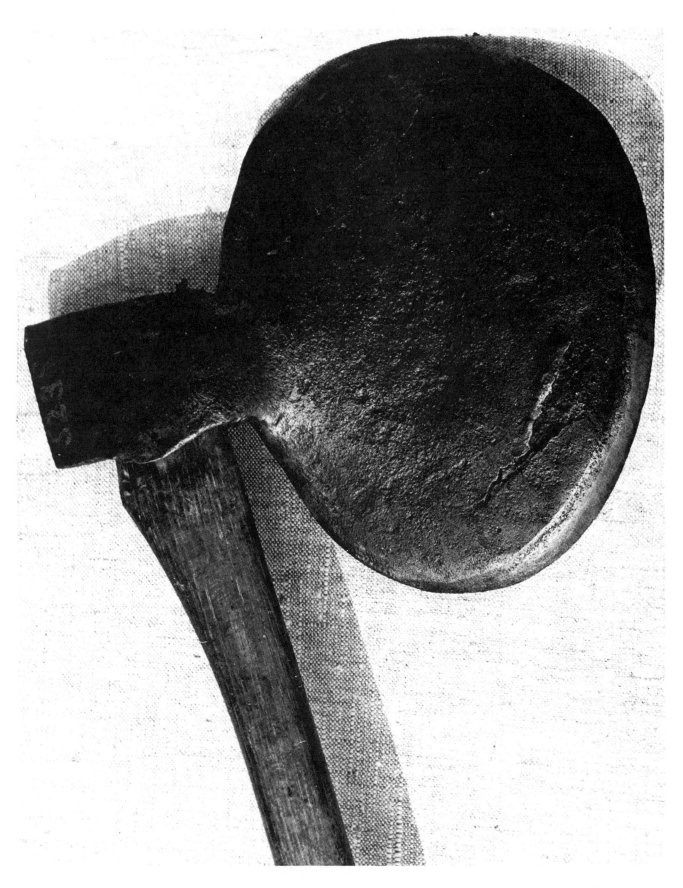

Believed to have been used for cutting turf, this circular axe is most unusual—but compare it with the example on Page 88. "C. J. Bishop" is possibly the original owner. (James A. Keillor collection)

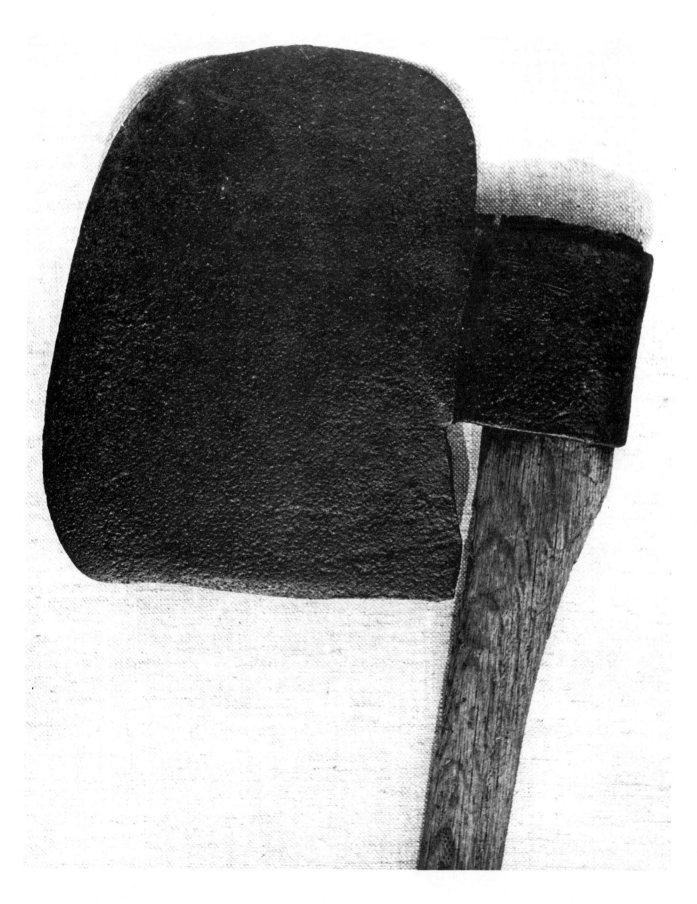

Also reportedly a turf axe, used in Massachusetts for cutting peat or sod for walls of cabins or other small buildings. (Author's collection)

NON-WOODWORKING AXES

A comparatively modern tool that harks back to the T-shape of the Middle Ages, this is called an "edging" axe by its owners, who believe it was used for trimming turf at the sides of drainage ditches. (Courtesy of Goschenhoppen Historians, Inc.)

The use of rivets in this axe is highly unusual, and bears out the report that it was used for cutting turf, not wood. (James A. Keillor collection)

NON-WOODWORKING AXES

Said to have been used for gashing the trunks of turpentine conifers, this axe's rounded bit—probably steel, as witness the difference in pitting on the poll—resembles that of an ice axe illustrated on Page 133. (James A. Keillor collection)

A mid-nineteenth-century ice axe. (James A. Keillor collection)

NON-WOODWORKING AXES

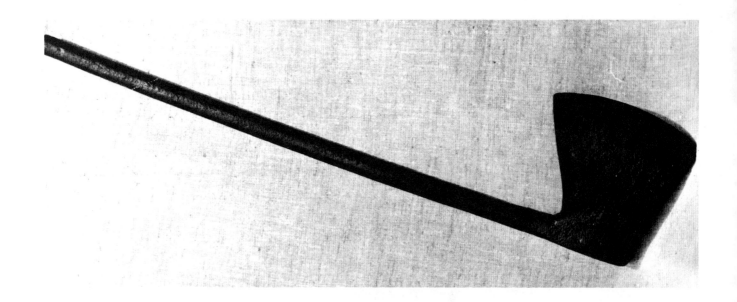

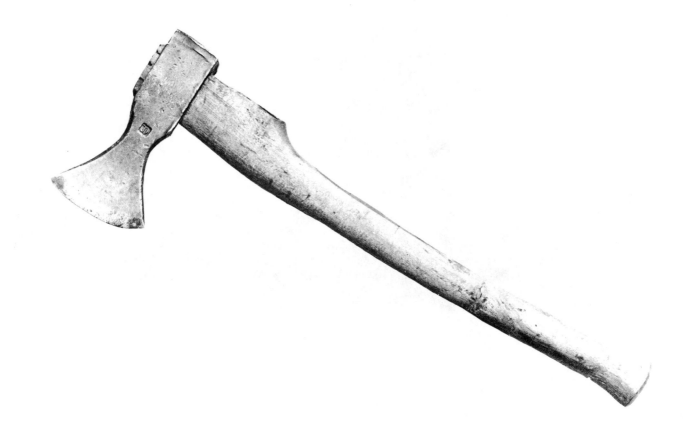

The small, highly specialized axe at top has an iron handle; uses, provenance, date, and method of manufacture are all unknown. Below is another "mystery" axe, possibly used for chopping caked, raw sugar from the barrel. (Above, the James A. Keillor collection; bottom, the John Grabb collection)

A bull axe made by Joseph Hoster of Philadelphia in 1840. The burred edge of the stud indicates that it was also used for a purpose other than felling cattle. (Author's collection)

NON-WOODWORKING AXES

An axe of unique shape and unknown function, with a 30-inch handle and a "laid" steel cutting edge. (Courtesy of the Mercer Museum)

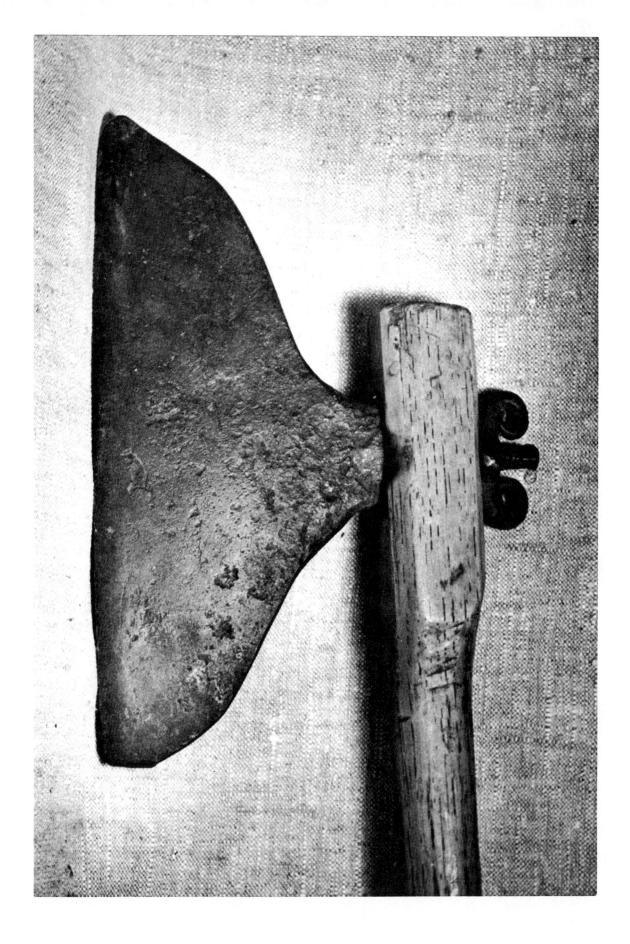

An antique form in a remarkable state of preservation. Fastening the axe to the handle with a wing-nut virtually eliminates it as a woodworking tool. (Courtesy of the Pennsylvania Farm Museum of Landis Valley)

NON-WOODWORKING AXES

V

Care of the Axe

The affluence of American society today and the plentiful supply of tools make it difficult for one to understand the importance of the axe in earlier times. However, the axe was as important to the pioneer as his gun: in fact, it was so highly regarded that both the maker and buyer were concerned with its original condition, as well as the service it rendered. If any imperfection was found which could be attributed to inferior material or workmanship, the tool was often replaced free of charge. Warranties (guarantees) were offered as early as the eighteenth century when John Ott advertised in the June 2, 1780, issue of the *New Jersey Gazette:*

Warranties

> The subscriber takes this method to inform the public, that he has a large quantity of the best German steel, and that he intends to apply himself wholly to making axes, in the neatest manner, which will be warranted.

Although the terms of his warranty are not spelled out in this advertisement, in a nineteenth-century advertisement of the Holmes Axe Factory (of Temperance Village, Pennsylvania) in *Harris' General Business Directory of the Cities of Pittsburgh and Allegheny,* printed by A. A. Anderson in 1841, the terms are specified:

> Where the following variety [Axes, Broad Axes, and Hand Axes] are manufactured, and warranted for thirty days if properly used, or if a scale or flaw in the steel or iron should injure them, the Axe may be returned, and it will be made good, or he will give another one in its place.

It is of particular interest to note that in the eighteenth and nineteenth centuries axe-makers were almost alone in offering warranties,

American Axes

but in the twentieth century, when many products are warranted, axes are not. Among several possible explanations, the most probable is that advanced techniques in quality control today minimize the chance that an imperfect axe will reach the market.

Manufacturers of the twentieth century, such as the Emerson and Stevens Manufacturing Company, Inc., of Oakland, Maine, stress the high quality of their products but stop short of a warranty. The following statement is taken from their broadside:

Quality claims

> Our axes are hammered out from rough bars, made with the cutting edge welded to the poll or soft body of steel which makes a strong sound job; carefully ground on wet grindstones; hardened and tempered one at a time by an experienced workman who has worked with us a long time; last of all they are carefully assorted, giving the user a good straight bit and eye that will hang on an axe handle correctly.

Most contemporary manufacturers imply by the description of their

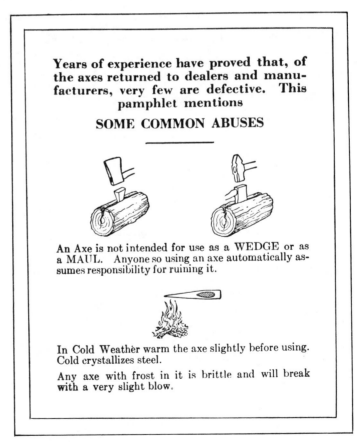

(AUTHOR'S COLLECTION)

Care of the Axe

products that their axes are equal to the finest made, but a curious statement appears in the recent advertising of one:

> GUARANTEE: All axes are guaranteed to be free from flaws or defects in workmanship or material, but are not warranted.

The following statement released by some of the most important axe manufacturers in the early twentieth century indicates their attitude toward quality and the extent of their responsibility:

> Abuse of Axes and The Care of Same: The return of axes for credit, when they are worn out or abused, is one of the greatest evils in the hardware trade, and we ask the assistance and cooperation of both wholesaler and retailer in our efforts to stop this abuse.
>
> We are manufacturers of many years experience, and we pledge our word and honor to our valued customers, that when we furnish first quality axes, they are made by expert workmen, of highest grade material, perfect temper, and free from flaws. The users should not expect replacement of axes with sound breaks, caused by foul blows, or of axes ruined by careless grinding or of axes used as mauls to drive wedges, battering the heads or breaking the eyes, or of axes the eyes of which have been broken by driving a metal or wooden wedge in the end of the handle too tightly with a heavy hammer or maul.
>
> The eyes of axes are made thin and symmetrical, and if the wedge is driven too tightly, the axe will be broken or bulged out of shape.
>
> When a dealer replaces an axe not defective, he does two things: viz., loses the sale of the second axe to which he is justly entitled; loses the profit on the first sale because the cost of handling the replacement eats up that profit. This vicious circle of loss once started, follows through to the jobber—and in turn to the manufacturers. Manufacturers suffer most because in addition to economic losses they sustain greater losses to their reputation, which has been built up through years of experience, hard work, and fair dealing.
>
> We know that it is hard for retailer or wholesaler to judge abused axes, and as a help to you, we have prepared and attach an educational pamphlet on axes, showing the usual abuses caused by improper usage or careless grinding. With this pamphlet you can show your customers when an axe has been abused, and can cut down the losses you have been assuming. If in doubt as to de-

The user's responsibility

American Axes

fect, *never replace* an axe, until it has been inspected by the manufacturer.

We, as manufacturers, desire to safeguard our reputation as quality makers, by furnishing in first quality axes, tools that will give service to the user.

As a measure of cooperation, we shall be glad to furnish additional copies of this letter and pamphlet for your salesmen or customers if you desire them.

Sincerely yours,

Kelly Axe & Tool Company
Fayette R. Plumb, Inc.
The Collins Company
Mann Edge Tool Company
North Wayne Tool Company
Warren Axe & Tool Company
Rexford Manufacturing Company
Emerson & Stevens Mfg. Co.

The installation of a handle has always been a matter of concern to axe makers and users. As the manufacturers' statement pointed out, the eyes are thin and symmetrical, and if a wedge is driven in too tightly, the axe will break or bulge out. One user suggests that the handle be heated before fitting it to the axe, which presumably shrinks the wood so that after installation it fits tightly forever.

It has been pointed out that, in the late nineteenth century, contour lathes were used to produce handles, and that machines also gave them a trial bend to be certain that they would not break after they were placed in an axe. A prime concern of twentieth-century manufacturers also has been the problem of holding the axe on the handle. An axe, flying off its handle, is obviously very dangerous, not to mention very inconvenient, when the woodcutter is miles away from the place where adjustments could be easily made. Therefore although the common wooden wedge served man for centuries, this simple device has been replaced by metal wedges and other devices. One company advertised a threaded wedge which could be turned in farther as the wood dried out, thus compensating for shrinkage of the handle. The advertising of Fayette R. Plumb Company, Inc., points out their superior method for fastening handles:

Fitting the handles

. . . Since striking tools were first used by primitive man, the

Care of the Axe

major problem has always been one concerning loose handles. Many patents have been issued through the years covering various methods and devices to keep handles tight or to retighten the handle after it has loosened. Virtually all of these patents are either impractical or clumsy. PLUMB PERMABOND has solved this. . . .

Plumb research developed PERMABOND to replace conventional wedging. PERMABOND is a specially formulated material that bonds together permanently two dissimilar products like steel and wood, steel and fiber-glass, etc. PERMABOND has advantages that no other type assembly can offer. Use these sales features below when selling Plumb Hammers, Hatchets, Axes, or Sledges.

The frontiersman achieved essentially the same result by sticking his axe in a bucket of water when he was not using it. Technology can be simple.

A woodcutter can keep his axe in a leather case, avoid hitting stones, and heat his axe in cold weather before he uses it, but in addition he must know how to grind it. A group of anonymous manufacturers has provided hints on how to do the job properly, including:

Grinding hints

1. Do not use a high-speed dry grinding wheel.

2. Grind slowly on a wheel kept very wet.

3. When re-grinding, start to grind from two to three inches back from the cutting edge and grind to about one-half inch from edge.

4. Work for fan-shape effect, leaving reinforcement at corners adequate for sufficient strength.

5. Strive for a convex bevel.

6. Be careful not to grind concave bevel. It will not offer enough strength behind the cutting edge and will break easily.

7. A long straight bevel is better than a concave bevel but still does not offer the strength of a convex.

American Axes

Although grinding is a very important part of sharpening an axe, it is only the first step. One expert points out that an axe needs grinding only one time; this is true assuming the axe is never nicked by striking a stone or other hard object.

Filing

After grinding, the next step is to file the edge. One axe manufacturer stocked a file, designed for this operation, which had coarse cutting teeth on one side, finer teeth for finishing the surface on the other. To sharpen the axe in the woods the poll is rested on the ground and the axe is supported in a slanting position by leaning it against a stone, the edge facing upward. The filing motion should be away from the edge for a distance of about three inches. Finally, a bevel should be filed about a half-inch wide along the cutting edge. First one side is filed and then the other. Some woodsmen, in too great a hurry, file only the bevel, but this procedure will create a stubby edge so that the axe will not sink into the wood. A double-bitted axe can be steadied by merely sinking one edge into a log to file the free edge, and then reversing the position of the tool for the second edge.

Honing

After filing, the axe should be honed. A hard fine-grit stone is best for this purpose. The stone, tilted forward slightly, is moved with a circular motion across the edge of the axe. The resulting burr is sheared away when honing is repeated on the other side. A residual burr will impair cutting and could induce abnormal deterioration along the edge. A honed axe will cut well and stay sharp.

VI

Use of the Axe

A survey of axe uses in the past reveals that an axe was employed for a variety of purposes such as felling, hewing and shaping, and mortising, and that each function needed a specific form. But basically the axe, in the minds of most people, has been a tool for cutting down trees and splitting wood. The felling axe, used for cutting down trees, has probably outnumbered all others; and because this axe could also be used for general utility purposes, most people are familiar with its shape.

There has always been a certain spirit of high adventure associated with cutting down a tree, and the following steps have developed over the years since colonial times, and remain the correct way today.

First, the natural "lean" of the tree is ascertained. This is done by standing some distance from the tree and holding the axe loosely in one hand, with its head hanging downward so the handle becomes a plumb line to gauge which way the tree inclines. The lean indicates the easiest direction in which to make it fall. Attention also must be given to the direction of the wind, particularly if the tree has leaves, as well as a clear spot for it to drop into, and the absence of entanglements as it falls. If no clear spot is available, the tree will have to be crashed into a dead tree that can be chopped away to release it, or into the nearest sound tree that will be least damaged by the fall. It is best to use the forces of nature if they fit the needs of the situation.

Felling a tree

To avoid the cross-graining from the root system, trees for lumber are cut high. This procedure is easiest for the chopper, as well as giving enough purchase for pulling or bulldozing the stump out later.

The first "box," or wedge-shaped cut, determines which way the tree will fall, unless otherwise guided by wedges, etc. The feller stands sideways to the tree to allow for the full swing of his sidearm motion; and,

American Axes

Notching

since he turns to face in the opposite direction to complete each of the two notches required to take down a large tree, he is (if a professional logger) able to chop right-handed or left-handed equally effectively. The bottom of the box is cut parallel with the ground, while its upper side slants downward into the horizontal cut at an acute angle. The first gashes are made shallow, so the axe is never completely imbedded in the wood. Moving inward, alternate horizontal and angular cuts are made, with the blade given a twist at the end of each blow so as to loosen the chip. The feller chops almost halfway through the tree to

Felling a tree: the chopper works from the left (on the slightly higher kerf), to make the tree fall to the right.

complete the notch which will guide its fall. The second notch—the one which will actually fell it—is started on the opposite side and a little higher than the first. The second one continues until only a narrow hinge of wood remains. Under normal conditions, this procedure fells the tree in the direction of the first notch.

Certain precautions are important in felling a tree. First, there must be room for a clear swing to be made with the axe; therefore any entangling small limbs or vines must first be cut away. Second, glancing blows are to be avoided, for the rebound of the axe can injure the chopper. Third, when the hinge breaks, the trunk can bounce backward and injure the feller or a bystander (felling a rotten tree is particularly dangerous, for, not having a strong hinge, it can fall in any direction). Fourth, flailing limbs have long been called "widow makers," so all personnel must stand outside their range as the tree crashes down. And fifth, it is important that the axe head be encased in a leather sheath when the tool is not in use, both to protect its edge *and* the person carrying the axe. Incidentally, to the writer's knowledge no such sheath from the eighteenth century has been found.

After the tree is on the ground, the limbs are easily trimmed off with an axe-stroke parallel to the tree trunk, working from the base toward the top. The limbs are cut smoothly and closely to facilitate skidding on to the rollway and thereby minimize transportation problems. Projecting stubs will complicate peeling the bark if this step is not part of the particular logging operation.

"Bucking" is the term used for cutting a tree into logs or bolts. Large trees are usually bucked in the woods to expedite their removal from adjoining trees, as well as to reduce the weight to be handled. The entire trunk is divided with a board-foot measuring bar, which ranges from short lengths for firewood and pulpwood to greater lengths for construction purposes. Thought must to be given to the division of a trunk so that crotches, a large knot, or deep blemishes can be avoided or removed when the log is cut to size.

In bucking the trunk of a large tree, the chopper stands on it and chops between his feet. The following excerpt from McLaren's *Axe Manual* describes the procedure:

> Always stand on top of any log that will give you a foothold. If the surface is smooth, roughen it with your axe to secure a firm foothold. Chop between your feet. Do not drive straight in: cut at an angle, usually about a 50-degree angle from the edge of the log. Many axe users fail to start their cut wide enough, and before they reach the center of the log, they are cutting in a tight angle and have to recut a wider notch. The width of the cut varies with the size of the log. On a log one foot in diameter, start your cut ten inches wide. On bigger logs your width increases so that a twenty-four-inch log would need about a nineteen-inch cut.

One of the most fascinating uses of the axe, but little practiced today, is hewing a square or rectangular shape from a round timber. There

Use of the Axe

Trimming

Bucking

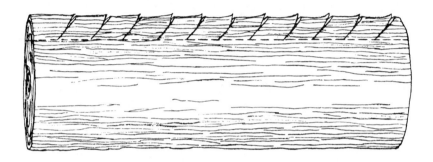

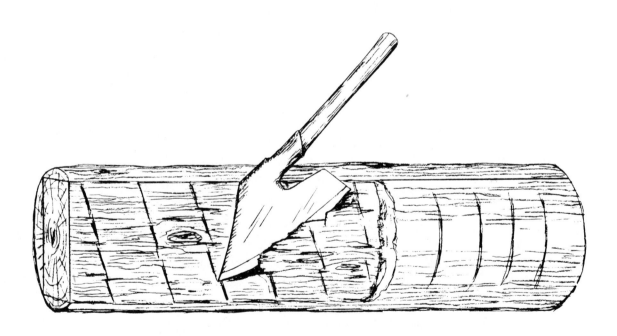

Three steps in squaring a timber: the log is held by dogs and a square marked on the smaller end; it is scored; then it is "hewed to the line."

were doubtless variations in the techniques used, but the general method was as follows.

Use of the Axe

First, the bark was removed. The log was placed crossways on two others to make the hewing comfortable and to avoiding striking the earth or stones with the axe. The log to be hewed was positioned near the ends of the two supporting logs, and held in place with a log-dog. There was a spur on each end of the dog, one being driven into the log to be hewed, the other into a supporting log. This arrangement kept the log from rolling. The size of the dog was determined by the size of the log, a big one requiring a long dog, etc.

Next, the largest square possible was laid out on the smaller end of the log, and if the log tapered sharply from one end to another, sometimes a square had to be laid out so large on the small end that the corners of the finished timber were rounded toward the small end. A square the same size as the first was laid out on the opposite end of the log. A notch was cut at the top of corresponding vertical lines on each end into which a string was pinched, covered with a marking compound such as charcoal. The line was then stretched taut and snapped on the log to make a line on the top from end to end which the hewer could follow. Then, standing on the log or ground, the hewer used a felling axe to cut gashes or notches in the tree, always stopping short of the line. If only small amounts of wood were to be removed, he made only small cuts.

Hewing

After scoring one side of the log, the hewer took up his broadaxe and, with a diagonal downward motion, "hewed to the line." Each cut slightly overlapped the previous one, and by skillful use of this ancient tool an amazingly smooth surface could be produced. Sometimes two hewers worked simultaneously on two sides of the log—an operation which created the need for left- and right-handed hewers. It is reported that a premium wage was paid to left-handers, because they were less plentiful. If timber was scarce and builders were not fastidious, logs were trimmed flat on only two sides. Many examples of such work can be found in large barns built in the nineteenth century.

Although it seems evident that some expertise was required to hew a log, the ultimate skill in the use of the axe probably was shown in building a log structure, since such construction required preparation of the logs and a complicated mode of making the corners.

The major problem in building a log structure was "corner-timbering," a technique used to secure the corner joints and thereby keep the wall erect and straight. In primitive buildings the logs were left round

American Axes

and a saddle type of joint was chopped near the ends of the logs with a felling or hand axe. A more sophisticated building was built of square logs neatly fitted with a dovetail joint in the corner, a method that was both attractive and long lasting.

For a building of round logs, the trees were felled, trimmed, and cut to length in the woodlot. The logs were taken to the construction site and laid out in four sections, each section to become the wall in which the logs were to be utilized. The heaviest logs were used at the bottom of the wall, the thinner ones near the top.

Corner-timbering First, four foundation timbers were carefully laid on cornerstones or foundations, with one skilled axeman working at each corner. Considerable dexterity was required of these men, for each had to cut a notch in the logs so that every log fitted snugly into the one previously laid. Some measurements were made, although the ends sticking out on each corner suggest that measurements were of a random sort. All the cutting was done "by eye" and a crooked corner quickly disqualified a man as an expert. As the walls became higher, slanting logs were rested on the top of the wall and additional ones were raised on forked sticks up to the axemen.

Further emphasis might be given here to the work done with an axe. A saddle notch was used on hastily built structures. This type of notch was a semicircular cavity cut partially through the log so the adjoining log rested firmly in a vertical position and could not be moved horizontally. Although some skill with an axe was required for this type,

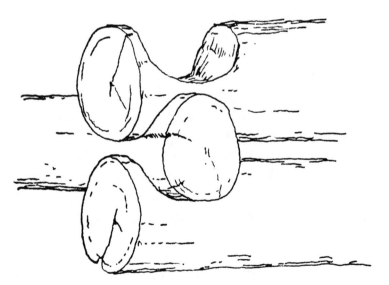

112 Corner-timbering: simple saddle notching.

the problem was simple when compared to constructing a dovetailed corner (discussed below).

Use of the Axe

Another technique used in corner-timbering round logs is known as V-notching. In this method a notch in the shape of a V was cut in the bottom side of each log, and the reverse shape was cut into the top of each log. This method required more time and skill than a simple saddle notch, but it does not seem to have been much of an improvement as far as function is concerned. If the logs were squared only on the top, the cut ends appeared to be shaped like a pear. If they were cut flat on the bottom, the cut ends looked like the gable end of a house.

The zenith of axe work in the construction of log buildings occurred in dovetail corner-timbering. In most cases each log was first hewed square before the corner joints were executed, a procedure which obviously required additional time and skill. The advantages of this method

Dovetailing

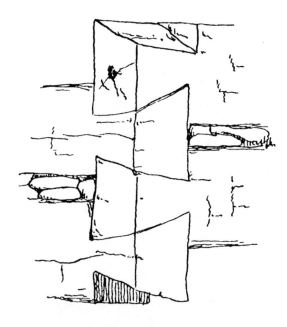

Dovetail corner-timbering, with chinking.

were that all surfaces cut on the corner of the log slanted downward, thus facilitating drainage and ensuring a subsequent long life for the building. It was an effective method for locking the timbers together, for it was virtually impossible for them to slip out of place. And, in most cases, the logs were brought very close together, thus requiring a small amount of chinking. It is also likely that the chinking rarely needed replacement, for it rested on a flat surface rather than a slanting one (as in the corner-timbering of round logs).

American Axes

Half-dovetailing was different from full-dovetailing in that the bottom notch was flat instead of slanting. It seems to have been a satisfactory procedure and was easier to effect. It is evident that considerable layout was required to form a dovetail joint for a log building. It also required very precise work with an axe, and an examination of the joints indicates that all the work was done with an axe.

Other modes of corner-timbering are known as diamond notch, half notch, double notch, half log, and mortise-and-tenon. The mortise-and-tenon method involved the fitting of the shaped ends of horizontal logs into cavities made in vertical corner logs, where they were kept in place by inserting a wooden pin. Excellent examples of this mortise-and-tenon work can also be seen in framed barns and in the trusswork of covered bridges.

There were doubtless many uses for axes other than those noted here. The axe was also used to split fenceposts and rails and to taper the rails so they could overlap in the postholes. A mortising axe was definitely used to remove rounded corners and the web left by the boring tool. These narrow axes are commonly known as "posthole axes."

VII

Directory of American Axe Makers

The purpose of this directory is to assist researchers and collectors to obtain information about the men who made axes in America from the eighteenth through the twentieth century. To the writer's knowledge this is the first time such a roster has been attempted, and it has been prepared with the hope that it will elicit further facts, and spark further interest in this hitherto generally overlooked facet of Americana. Subsequent research will doubtless bring other names to light, as well as enlarge on the data about known makers, even though perfect documentation for all makers can never be achieved.

Information varies with each entry. For some, only a working date or location is given. Although this situation is unfortunate, it was obviously unwise to delete the name of any maker on the premise that complete data is not now available: certainly every collector is grateful for each fragment of information about a maker in whom he is interested; and of course any small bit of information could lead to complete identification on the scene where the maker worked. For similar reasons, amplifying comment in italics has been appended to various entries.

Entry key

The date assigned to the entries indicates when the craftsman or craftsmen worked; or, in some cases, public notice of when a manufactory was in operation; or when a patent was obtained. Data about the latter is quite extensive in the U.S. Patent Office. Obviously, complete information such as this could not be supplied in the present directory, but by supplying a patent number to the U.S. Patent Office, Alexandria, Virginia, a complete copy of the patent can be obtained for a small fee.

American Axes

Adams, Peter : Blacksmith, Lancaster, Pennsylvania. 1843

Allen, J., Sr. : Pise, Indiana. Patent (mode of making axes) April 9, 1829

Althouse, Isaac : Axe-maker, Columbus, Ohio. 1875

American Axe & Tool Co. : Axe-makers, Cleveland, Ohio. 1894

American Fork & Hoe Co. : Charleston, West Virginia. [*Parent company of Kelly Axe & Tool Works,* q.v.]

Anderson, James : Blacksmith, Williamsburg, Virginia. 1780
[Letter: James Anderson to George Muter, July 28, 1780: *Sir, In my present situation, I think I can make eighteen axes and three tomahawks a day. I am Sir, your very humble servant to command. James Anderson*]

Andrews, J. : Dinwiddie County, Virginia. Patent (vertical-step or gudgeon axe) August 11, 1817

Angell, E. : Axe manufacture, Newport, New York. 1850

Arnold, Ephraim : Edge-tool manufacturer, New York, New York. 1866

Atkinson, Jos. : Edge-tool maker, Buxton, Maine. 1829–1832

Avery, Samuel : Axe and plow maker, Charlemont, Massachusetts.

Bailey, _____ : Blacksmith, Bridgton, Maine. 1816–1832

Bailey, D. E. : Patent #1,536,872. 1925

Baird, Samuel : Axe manufacturer, Limestone, Pennsylvania. 1870–1871

Baker, Holmes & Brown : Edge tools, Baltimore. 1857

Bakewell, _____ : Patent (design for an axe) #16,631.

Barbour, J. : Patent (design for an axe) 1896

Barnett, Thomas : Edge-tool manufacturer, Philadelphia. 1868

Bartholomew, E., Tool Company : Mill Hall, Pennsylvania. Patent (axe-bit blank machine) #16,011. 1871

Barton, D. R., Tool Company : Rochester, New York.

Beardsby & Co. : Axe manufacturer, Ontario County, New York. 1850

Beatty, John C. : Edge-tool maker, Chester, Pennsylvania. 1874

Beatty, John C. & Bros. : Edge-tool maker, Springfield, Pennsylvania. 1870

Beatty, Wm. & Sons : Edge-tool manufacturer, Philadelphia. 1864

Beck, B. : Patent (for an axe) #464,910. 1891

Beck, Wm. : Edge-tool manufacturer, Philadelphia. 1868

Beidler, J. H. : Adrian, Michigan. Patent #93,585. 1869

Benjamin, J. S. : Axe manufacturer, Dayton, Ohio. 1868

Benjamin, Timothy : Edge-tool maker, New York City. 1826–29

Bentley & Craddock : Edge-tool manufacturer, Philadelphia. 1868

Benton, _____ : Patent (axe poll) #812,403. 1906

Betterdsley & Co. : Axe manufacturer, Malone, New York. 1850

Bieser, William : Edge-tool manufacturer, Philadelphia. 1876

Bill, H. N. & J. C. : Willimantic, Connecticut. Patent (method of securing helves in axes) #11,350. 1854

Billings, Jno. : Edge-tool maker, Saco, Maine. 1825

Binkley, John Z. : Edge-tool maker. 1875

Black, _____ : Patent (axe or wedge for splitting wood) #427,088. 1890

Directory of American Axe Makers

Blair, C. : Manufacturer of axe bits, Collinsville, Connecticut. Patent #93,066. 1869

Blaisdel, H. : Axe manufacturer, Concord, New York. 1850

Blaisdell, R. : Axe manufacturer, Springville, New York. 1850

Blake & Carpenter : Scranton, Pennsylvania. Patent (axe-poll swaging machine) #89,623. 1869

Blanchaw, _____ : Axe manufacturer, Homer, New York. 1870

Blood, Isaiah : Edge-tool manufacturer, New York City. 1866

Blount, _____ : Patent #747,783.

Booth & Mills : Edge-tool manufacturer, Philadelphia. 1868

Boyle, Aaron : Brandon, Ohio. 1853

Brady Edge Tool Works : Lancaster, Pennsylvania. 1883

Brady, W. : Edge-tool maker, Lancaster, Pennsylvania. 1875

Brady, William & Son : Edge-tool makers, Mt. Joy, Lancaster County, Pennsylvania. 1869

Breckenridge Tool Co. : Axe manufacturers, Toledo, Ohio. 1883

Broadaxe made by Brady of Lancaster, Pennsylvania. (Courtesy of the Pennsylvania Farm Museum of Landis Valley)

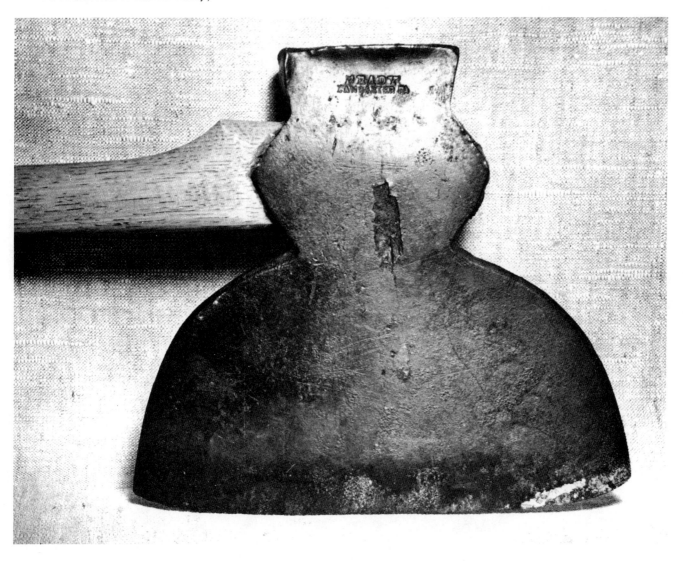

American Axes

Bringhurst & Co. : Edge-tool manufacturer, Philadelphia. 1866

Brooks, Brewster : Axe manufacturer, Stony Point, New York. 1870

Brown, S. J. : Axe manufacturer, Berea, Ohio. 1863

Buckley, _____ : Patent (combination tool) #766,808. 1904

Bunnell, J. N. : Patent (axe and ice-pick) #109,867. 1870

Bunton, W. : Pittsburgh, Pennsylvania. Patent (axe-poll blank) #95,646. 1869

Butter, B. : St. Johnsbury Center, Connecticut. Patent (axe-handle shield) #86,130. 1869

Carter, I. M. : Hyde Park, New York.

Carter, Pulaski : Axe manufacturer, Scranton, Pennsylvania. 1841–1870

Carver, _____ : Patent for an axe. 1891

Chaplin, Freeman : Axe manufacturer, Rochester, Pennsylvania. 1870

Chapman, L. : Collinsville, Connecticut. Patent (machine for forming axe bits) #90,726. 1869

Clark, David : Blacksmith with triphammer, Northfield and Franklin, New Hampshire. 1824

Cobb, William : Axe manufacturer, Rochester, New York. 1816

Cochran, George : Hardware merchant, Pittsburgh, Pennsylvania. 1844

Cohoes Axe Mfg. Co. : Cohoes, New York. 1880

Colburn, D. W. : Patent #66,563. 1869

Colburns & Fenn : Ansonia, Connecticut. 1856

Colchester, Daniel : Edge-tool manufacturer, Altoona, Pennsylvania. 1870

Collins, A. : Winsted, Connecticut. Patent for an oval axe.

Collins & Co. : Edge-tool manufacturer, New York City. 1866

Collins Company, The : Edge-tool manufacturer, Collinsville, Connecticut. 1826–1966

Collins, David : Axe-maker, Hartford, Connecticut. 1825

Collins, Samuel : Axe manufacturer, Canton, Connecticut. 1828

Cox, H. E. : Patent (reversible axe) #1,264,776. 1918

Craddock, Thomas : Edge-tool manufacturer, Lockport, New York. 1826 [From the *Lockport Observatory*, 1826: *Thomas Craddock having commenced the above business at the sign of the Broad Axe the corner Main and Transit Sts. nearly opposite the Washington House in the Village, respectfully informs the inhabitants of this vicinity and the country at large that he manufactures Edge Tools of almost any description wanted in the country which he will sell at wholesale and retail and warrant of superior quality. N.B. Strict attention paid to all branches of country work and most kinds received in payment.*]

Crossby, C. and Sons : Edge-tool manufacturer, Philadelphia. 1868

Curil, F. C. : Lancaster, Pennsylvania. Patent (manufacture of axes and hammers) #82,607. 1868

[Dates Patent Steel Company : Axe manufacturer, Toronto, Canada. 1876]

Davis, N. : Axe manufacturer, Wawarsing, New York. 1850

DeWitt, Morrison : Edge-tool manufacturer, Philadelphia. 1868

Didero, J. M. : Patent (design for an axe) #20,814.

Dodge, L. : Cohoes, New York. Patent (shears for the manufacture of axes) #31,927. 1861

One of the main buildings of the Collins Company of Collinsville, Connecticut, as it appeared shortly after the company was bought by the Mann Edge Tool Company in 1966. (Author's collection)

Dodge, L. : Waterford, New York. Patent (machinery for hammering heads of axes) #54,311. 1866

Douglas Axe Company : Axe manufacturer, Boston [*Exhibitor at Centennial Exhibition*]. 1876

Douglas Axe Company : Edge-tool manufacturer, Boston. 1866

Douglas Axe Manufacturing Company : New York City. 1866

Douglas, Norton : Axe manufacturer, Watertown, New York. 1850

Duck, Hiram : Edge-tool maker, Lancaster County, Pennsylvania. 1875

Dunlop and Madeira : Edge-tool maker, Chambersburg, Pennsylvania. 1837

120 A broadaxe made by Dunlop and Madeira of Chambersburg, Pennsylvania. (Author's collection)

[From *Recollections of Chambersburg, 1830–1900*, by Cooper: *Next to the Paper Mill, the Edge Tool Factory was the most important manufacturing establishment. It was located where Sierer's manufacturing long occupied the ground and utilized the water power, and was carried on by Dunlop and Madeira. Mr. Dunlop resided on what has for sixty years been the Kennedy farm below town, and I believe the grinding department of the factory was first located there, but was subsequently annexed to the concern in town. . . . The Edge Tool Factory, for some years after I first saw it, was the busiest hive in town. It ran until late at night, and . . . the strong breath of the bellows made the fires roar and the hammers of the workmen made the anvils ring as they pounded the iron and steel into the shapes of axes. Dunlop and Madeira deserved to make a fortune, but I believe they did not. George A. Madeira was an amiable and intelligent man, but perhaps lacked the rugged and relentless energy required to bring fortune from adverse circumstances. Times were hard in 1837 and continued so for many years, and then Pittsburgh and other westward points came rapidly forward with manufactures of iron and cut off trade of our "Lemnos Factory," a name borrowed from Greece and suggestive of interest in that country's struggle for Liberty.*]

Dunn Edge Tool Company : Oakland, Maine.

Early, W. & B. : Axe manufacturer, Malone, New York. 1850

Ehret, J. & Son : Edge-tool manufacturer, Philadelphia. 1876

Ellis, J. W. : Pittsburgh, Pennsylvania. Patent (method of making axe blanks) #68,423. 1867

Emerson, J. E. : San Francisco, California. Patent (axe-handle fastening) #27,784. 1860

Emerson & Stevens : Edge-tool manufacturer, Oakland, Maine. 1870–1965

Empire Tool Works : Axe manufacturer, Cohoes, New York. 1876

Empire Works : Edge-tool manufacturer, Pittsburgh, Pennsylvania. 1857

Estep, D. P. : Pittsburgh, Pennsylvania. Patent (axe-poll making) #15,880. 1856

Estep, Ephraim : Edge-tool manufacturer, Lawrenceville, Pennsylvania. 1837
[*"Employs about 30 hands, and makes about ten dozen of cast steel axes per day, broad axes, hatchets, etc. Annual amount about $60,000."*]

Estep & Sons : Manufacturer of edge-tools, shovels, etc., Lawrenceville, Pennsylvania. 1844
[*"Axes, hatchets and all kinds of carpenter and cooper tools, made in the best style, and warranted of a superior quality constantly on hand at the works, or at the warehouse of Mr. George Cochran, No. 25 Wood Street, Pittsburgh (Pennsylvania). P.S. Extra large knives and any article in our line made to order at short notice. Orders addressed to the Subscribers or to George Cochran promptly attended to."*]

Everett, C. O., Saw Co. : Bangor, Maine.

Everson, _____ : Patent (axe) #352,972. 1866

Ewald, John W. : Edge-tool manufacturer, Baltimore, Maryland. 1857

Fenn, Leonard : Axe-maker, Ansonia, Connecticut. 1856
[From Colburns & Fenn broadside:

Directory of American Axe Makers

American Axes

Leonard Fenn, having had more than thirty years experience in the business of manufacturing Cast Steel Axes, expressly for the Retail Trade, would offer them to the public, having adopted a new and improved mode of tempering, giving to the steel a solid and fine edge, one that will stand longer without grinding, giving to them greater durability than any heretofore made.]

Fernald, Donald : Edge-tool maker, Saco, Maine. 1816–1832
["*In 1832 Mr. Fernald used one chaldron sea coal and 1,400 bushels charcoal.*"]

Forest City Tool Company : Cleveland, Ohio. 1861

Fowler, T. : Axe manufacturer, Russell, New York. 1850

Fowler, T. : Axe manufacturer, St. Lawrence County, New York. 1850

Francis Axe Company : Axe manufacturer, Buffalo, New York. 1907

Franklin, J. : Springfield, Ohio. Patent #981,416. 1869

French, Isaac : Axe manufacturer, Ravenna, Ohio. 1853

Frost, Benjamin : Blacksmith, Westbrook, Maine. 1827

Garlick, et al. : Patent (axe and cutter) #618,658. 1899

Gillingham, Wm. : Edge-tool maker, Baltimore, Maryland. 1831

Goodell & Waters : Edge-tool manufacturer, Philadelphia. 1876

Graeff, Joseph & Co. : Axe manufacturer, Allegheny, Pennsylvania. 1870

Graham, _____ : Patent (axe blade) #653,234. 1900

Griffin, Jones : Edge-tool manufacturer, Cleveland, Ohio. 1866

Grimes, W. A. : Patent #25,154. 1896

Hadcock, Solomon : Waterlong, New York. 1850

Hagen, Elijah : Axe-maker, Martic Township, Lancaster County, Pennsylvania. 1850

Hager, David : Edge-tool maker, Mt. Nebo, Lancaster County, Pennsylvania. 1875

Hagerman, N. : Axe manufacturer, Concord, New York. 1850

Hall, _____ : Blacksmith, Norway, Maine. 1832

Hammond Edge Tools : Philadelphia. 1864

Hammond, G. & Son : Edge-tool manufacturer, Philadelphia. 1876

Hannum, Caleb : Axe-maker, Norwich, Massachusetts.

Hannum, Josiah : Axe-maker, Williamsburg, Massachusetts. 1820

Harrington, Lewson : Blacksmith, Westborough, Massachusetts. 1832

Harrison, Alexander : Axe manufacturer, New Haven, Connecticut. 1834
[From *A History of American Manufacturers from 1618 to 1860*, by J. Leander Bishop: *American axes and locks were acknowledged to be the best in the world. There were two axe factories at New Haven, Connecticut, those of Alexander Harrison, and of the Collins Company; the latter was capable of finishing two hundred axes per diem, the former fifty. The steam factory of Mr. Maule, twelve miles from Wheeling, Virginia manufactured to the value of $10,000 per annum. Door locks began to be made there the next year by Pierpont and Hotchkiss.*]

Harrow, Isaac : Planing and blade mill, Trenton, New Jersey. 1734
[Notice in *American Weekly Mercury* for September 5–12, 1734:

Lately set up at Trenton in New Jersey, a planing and blade mill, by Isaac Harrow, an English smith, who makes the undernamed goods, viz. Dripping pans, Garden tools, Bark shaves, Glovers shears, Food Knives, Frying pans, Common shovels, Pot ladles, Scythes, Ditching shovels, Chaffing dishes, Peel shovels, Melting ladles, Mill saws, Broad axes, Coopers axes, Fire-shovel pans, Cross-cut saws, Carpenters tools, Smoothing irons, Cloathiers sheers, Coffee roasters, Coopers tools, Cow bells, Garden sheers, Hay knives.]

Hartman, John : Edge-tool manufacturer, Cleveland, Ohio. 1872

Harvey, _____ : Axe-maker. Patent #327,164.

Hazard, J. P. : Maker of axes, scythes, hoes, forks, etc., Washington County, Rhode Island. 1833

Hewes Edge Tool Company : Lancaster County, Pennsylvania.
[From *History of Lancaster County* by Ellis & Evans: *Peter's Creek comes in from Drumore, runs first a southerly course to Wick's Mill, then rather southwesterly till it meets the Puddle Duck, which, rising near the middle of the township (Fulton), runs in a winding westerly course, passing and giving power to George Hewes' edge tool factory.*]

Hight, Amos : Blacksmith, Scarborough, Maine. 1832

Hight, George : Edge-tool maker, Gorham, Maine. 1815

Higley, M. E. : Edge-tool maker, Cleveland, Ohio. 1861

Hilton, J. W., and Greene, R. W. : Edge-tool makers, Bradford, Pennsylvania. Patent #81,635. 1868

Hinman, D. : Canton, Connecticut. 1833

Hinman, D. : Winchester, Connecticut. Patent for an axe hatchet. 1833

Hitchock, Levi : Edge-tool maker, Williamsburg, Massachusetts. 1829

Hobbs, J. : Blacksmith, Norway, Maine. 1832
[From a company report: *In 1832 Mr. Hobbs used 2 tons Russian iron, and 200 pounds of Swedes blistered steel.*]

Hodgen, Cobb : Edge-tool maker, Gorham, Maine. 1832

Holmes, Alfred & Griffin, John & Son : Blacksmiths, York County, Maine. 1832

Holmes and Essington : Axe manufacturers, Milesburg, Pennsylvania. 1870

Holmes Axe Factory : Temperance Village, near Pittsburgh, Pennsylvania. 1841
[From a contemporary warranty: *Where the following variety are manufactured, and warranted for thirty days, if properly used, or if a scale or flaw in the steel or iron should injure them, the axe may be returned, and it will be made good, or he will give another one in its place. List of articles as follows: Axes, broad axes, hand axes, hatchets, foot adzes, and all kinds of Coopers tools, the above made of the best refined extra cast steel.*]

Holmes, John : Edge-tool manufacturer, Lockport, New York. 1850

Holt, L. B. : Adar Falls, Iowa. Patent (axe-handle guard) #105,235. 1870

Horn & Ellis : Edge-tool manufacturer, Philadelphia. 1868

Horton & Arnold : Edge-tool manufacturers, New York City. 1866

Hoster, Joseph : Axe-maker with shop "rear 285 Vine," Philadelphia. 1840

Directory of American Axe Makers

American Axes

Hubbard, ——— : Patent (axe) #463,002. 1891

Hubbard & Blake : Axe manufacturers, W. Waterville, Maine. 1907

Huber & Co. : Edge-tool manufacturer, Chambersburg, Pennsylvania. 1870

Huber, Henry : Edge tools, Philadelphia. 1864

Huber Tool Works : Philadelphia. 1876

Hudgens, Davis & Co. : Axe-makers, Baltimore, Maryland. 1857

Hufford, ——— : Patent (axe head) #424,205. 1890

Humphreysville Mfg. Co. : Edge-tool maker, New York City. 1866

Hunt, Warren : East Douglas, Massachusetts. Patent (axe-testing machine) #15,656. 1856

Hunter, W. A. : Axe manufacturer, Cleveland, Ohio. 1872

Huntermark, L. : Edge tools, Cleveland, Ohio. 1866

Hunt's Superior Axes : Axe manufacturers (Douglas Axe Mfg. Co.), East Douglas, Massachusetts. 1907

Hurd, E. F. : Axe manufacturer, Johnsonville, New York. 1907

Hurd, E. F. : Hoosick Falls, New York. Patent (axe-making machine) #121,172. 1871

Hurman, Nathan : Edge-tool maker, Buxton, Maine. 1828

Hutchins, C. : East Douglas, Massachusetts. Patent (axe-making machine) #16,732. 1857

Hyde, S. : Williamsburg, Massachusetts. Patent for an oval axe. 1830

Jamestown Axe Co. : Jamestown, New York. 1907

Jay, C. : Mill Hall, Pennsylvania. 1871 [*Trademark of a subsidiary brand of axe from the Mann Edge Tool Co. q.v.*]

Jenkins & Tongue : Edge-tool manufacturers, Philadelphia. 1868

Johnsonville Axe Mfg. Co. : Johnsonville, New York. 1907

Johnsonville Axe Mfg. Co. : Pittstown, New York. 1859

Jones, Edward : Edge tools, Philadelphia. 1864

Jumper, A. H. : Summon, Indiana. Patent #94,318. 1869

Karns, J. H. : Patent #561,000. 1896

Kelley, J. P. : Patent #520,738. 1894

Kelley, J. P. : Patent (axe or hatchet) #29,680. 1898

Kelley, J. P. : Patent (design for an axe) #26,513. 1897

Kelly Axe Manufacturing Co. : Louisville, Kentucky. 1889

Kelly Axe & Tool Works : Charleston, West Virginia. [*Maker of "W. C. Kelly Perfect Axe"; subsidiary of American Fork & Hoe Co.*, q.v.]

Kelly, W. C. : Patent (axe) #327,275. 1885

Kelly, W. C. : Patent (axe) #402,936. 1889

Kelly, W. C. : Patentee of "Perfect Axe," Louisville, Kentucky. 1889 [See also *Kelly Axe Manufacturing Co.*]

[From *William Kelly: A True History of the So-Called Bessemer Process,* James Newton Boucher: *Mr. Kelly began his business career when a young man before the elder Kelly had reaped any advantages worth considering from his discovery. He invented a process for making steel similar to the Bessemer invention. He therefore received no financial assistance from his father. He began the manufacture of edge tools in Louisville, Kentucky, in a room about thirty by thirty feet, with only a few employees to assist*

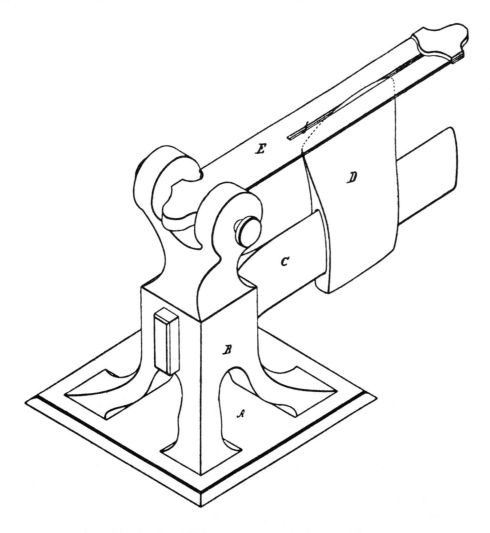

UNITED STATES PATENT OFFICE.

WARREN HUNT, OF EAST DOUGLASS, MASSACHUSETTS.

MACHINE FOR TESTING AXES.

Specification of Letters Patent No. 15,656, dated September 2, 1856.

To all whom it may concern:

Be it known that I, WARREN HUNT, of East Douglass, in the county of Worcester and State of Massachusetts, have invented a new and useful Machine or Tool for the Purpose of Testing the Trueness of Axes as They are Forged or Ground, of which the following is a full, clear, and exact description, reference being had to the accompanying drawing, in which is represented a perspective view of my machine.

It is very essential to the perfect operation of an ax that its cutting edge be in a plane passing through the axis of the helve, and my invention consists of a simple and effective tool, by the use of which the trueness of axes as they are forged or ground may be tested by the workmen as they proceed.

To enable others skilled in the art to understand my invention I will proceed to describe the manner in which I have carried it out.

In the accompanying drawing A, is the base to which is attached or cast the standard B. To this standard is secured a metallic bar C which tapers slightly toward its end and is of a size calculated to fill the eye of the ax D, and hold it motionless. E, is a gage plate which pivots in the top of the standards and may be raised and lowered in a plane passing through the axis of the bar C.

f is a slot made through the gage plate directly over the center of the bar C.

The operation of this tool is as follows:

The ax is put upon the bar C. and is slipped up toward the standard until it fits tight, the gage plate is then allowed to descend upon the edge, when by placing the eye over the slot *f*, the smallest variation from true may be detected in the edge, the workman having one of these tools by his side tests each ax as it is forged or ground in the most expeditious manner, and far more perfectly than can be done by the means heretofore employed.

What I claim as my invention and desire to secure by Letters Patent, is—

The within described tool for testing the trueness of axes consisting essentially of the bar C, and slotted gage plate E, operating in the manner substantially as set forth.

WARREN HUNT.

Witnesses:
EDWIN MOORE,
OLIVERE HUNT.

Drawing and specifications for Warren Hunt's 1856 axe-testing-machine patent. (Courtesy of the United States Patent Office)

American Axes

in the work. Like many other American factories, through his efforts and good business management, his works grew from year to year, until now in Charleston, West Virginia, he is the owner of the largest edge tool factory in the world. It is larger than all other factories of its kind in the United States put together, and it is in both capacity and output, than all the other similar factories in the world combined. His products are shipped partly in his boats, to nearly all parts of the world, namely, England, Germany, Russia, Australia, to most of the countries of South America, and to all parts of the United States and Canada. They are sold under the trade mark of "W. C. Kelly" his name alone being sufficient guarantee of their high quality, in whatever land they are offered for sale. In other states he is the builder and owner of woodworking factories which make handles for his tools and for sale to other factories and to foreign nations.]

Kidd, Charles, Cast Steel Edge Tool Factory : "Between Green & Penn Sts.," Baltimore, Maryland. 1831
["*Where may be found a good assortment of ships carpenters' axes, adzes, chisels, etc., House carpenters' hand axes, hatchets, chisels, etc., Mill Wrights' broad axes, adzes, chisels, gouges, etc., Coopers' axes, adzes, drawing, rounding, and hollowing knives, etc., Block makers' round axes, gouges, and chisels, Masons' hammers, picks, Mattocks, grubbing hoes, stone sledges, and quarry tools. All kinds of millwork made and repaired in the best manner. Orders for axes, tools, or iron work of any kind thankfully received and promptly attended to. All persons in want of tools, will do well to call and judge for themselves.*"]

King, _____ : Patent (axe head) #700,780. 1902

King, _____ : Patent (wood-chopping axe) #427,088. 1890

King Axe Company : Cleveland, Ohio. 1894

King Axe and Tool Company : Oakland, Maine.

King, Collins : Edge tools, Philadelphia. 1864

Kinnel, Philip : Edge tools, Columbus, Ohio. 1866
[From *Review of Columbus, Ohio*, 1887: *Mr. Kinnel established his business in this city on a rather small scale in 1866, but owing to his superior workmanship and strict attention to business, his trade increased and expanded year by year, until it reached its present immense proportions extending to all parts of the United States. This establishment is located at the corner of Mound and Front Streets, where upwards of ten first class mechanics are employed. Mr. Kinnel is a thorough mechanic, and personally superintends every department of his work. He manufactures edge tools of every description, using none but the best wares obtainable. A few of his specialties are double steel axes, iron and wood-handled cleavers, hedge knives, and mill picks.*]

Kloman, Bickerise, & Co. : Edge-tool manufacturers, Pittsburgh. 1870

Knapp, H. : Axe-maker, Cincinnati, Ohio. 1831

Knight, Sam. P. and others : Axe-makers, Cornish, Maine. 1832

Krause, George : Hardware company, Lebanon, Pennsylvania.

Lake Erie Tool Co. : Cleveland, Ohio.

The name "Perfect Axe" etched on the surface of this American felling axe represents the zenith in advertising as far as the author is concerned: presumably the design was beyond improvement. The rib and the hollows are its major qualities *(see also Page 74)*. (From a catalogue courtesy of Mann Edge Tool Company)

The claim is spelled out in lettering etched on the reverse side of the axe itself:

"This axe is made of the finest steel and is hand-hammered, tempered, and tested before leaving the shop. The blade is so shaped it will cut deeper but will not bind. It will burst the chip and it will not become stubbed. It has a tapered eye which binds hard. Try it and you will use no other." (Author's collection)

American Axes

Lame, C. H. : Edge tools, Philadelphia. 1864

Larew, George : Edge-tool maker, Sugartown, Pennsylvania. 1875

Larne, George : Gochenville, Pennsylvania. 1874

Lauriat, J. B. : Axe manufacturer, Little Falls, New York. 1850

Lawton, _____ : Patent (axe or hatchet) #712,554. 1902

Leather, _____ : Patent (axe) #755,890. 1904

Leverett, John : Axe manufacturer, New York City. 1859

Leverett, Josiah : Edge-tool manufacturer, New York City. 1866

Lewis, J. L. : Pittsburgh. Patent (making axes) #53,155. 1866

Libby, David : Edge-tool maker, Buxton, Maine. 1827

Lippincott & Co. : Pittsburgh. 1850 *["Lippincott & Co. manufacturers of hammered and cast steel shovels and spades, axes, hatchets, Muley mill cut and circular saws, forks, hoes, picks, mattocks, etc."]*

Lippincott & Co. : Pittsburgh. 1860–1907

Lippincott, J. : Pittsburgh. Patent (making axes) #27,227. 1860

Lippincott, J. : Pittsburgh. Patent (manufacture of axes) #53,155, 1866; patent (axe-bit blank bar) #85,110, 1868.

Little, Charles : Edge-tool manufacturer, New York City. 1866 *["Charles S. Little & Co., No. 200 Broadway. Tools—Unsurpassed in quality and variety, embracing the most approved Foreign and Domestic Manufacture, and adapted for the use of Coopers, House Carpenters, Ship Carpenters, Coach Makers, Millwrights, Machinists, Founders, Smiths, Miners, Tanners, Curriers, Etc."]*

Livingston, W. H. : New York City. Patent (fastening axes to handles) #30,146. 1860

Lockport Edge Tool Company : Lockport, New York. 1860 [See also *Simmons, Daniel*]

Love, I. E. : Patent (folding axe) #1,-515,688. 1924

McGregor, _____ : Axe manufacturer, Ohio City, Ohio. 1853

The etched "label" of the George Krause Hardware Company of Lebanon, Pennsylvania. Stampings of this sort are quite rare. (Courtesy of the Pennsylvania Farm Museum of Landis Valley)

Directory of American Axe Makers

McIlwain Bros. : Philadelphia. 1876

McIntosh Hardware Co., The : Axes and edge tools.

McKeever, S. & Son : Edge-tool manufacturer, West Grove, Pennsylvania. 1870

McKoy, ——— : Patent (turpentine gatherer's axe) #869,381. 1907

Mackey, J. : Napanock, New York. Patent (axe-dressing machinery) #5,736. 1848

Mackie, J. F. : New York City. Patent for making axes, hoes, scythes, pitchforks, etc. 1833

Mann Edge Tool Company, Axe manufacturer, Lewistown, Pennsylvania.

Mann, Harvey : Bellefonte, Pennsylvania. Patent (axe grinding) #93,727. 1869

Mann, J. Fearon : Axe manufacturer, near Bellefonte, Pennsylvania. 1907

Mann, J. R. : Patent (edge tool) #561,409. 1896

Mann, Robert : Axe manufacturer, Mill Hall, Pennsylvania. 1870 [See also *Jay, C.*]

Mann, Robert & Sons : Mill Hall, Pennsylvania. 1888

Mann, Stephen : Axe manufacturer, East Tennessee, New York. 1840

Mann, William & Co. : Axe manufacturer, Reedsville, Pennsylvania. 1870

Mann, William, Jr., & Co. : Makers of axes and edge tools, Lewistown, Pennsylvania. 1882

Marble, W. L. : Patent (safety guard) #604,624. 1898

Marsh Axe & Tool Company : Oakland, Maine. 1925

Marshall, Joel : Edge-tool maker, Buxton, Maine. 1807

Matthews, Pitt : Axe manufacturer, Jefferson County, New York. 1850

Maule, ——— : Axe manufacturer,

Drawing showing axes—without handles—packed for shipment. Customers would "hang" the axes themselves. (From a catalogue of the Mann Edge Tool Company)

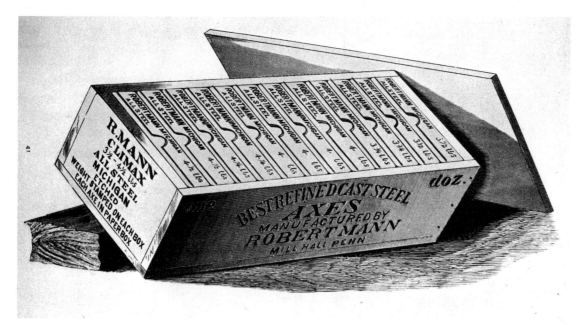

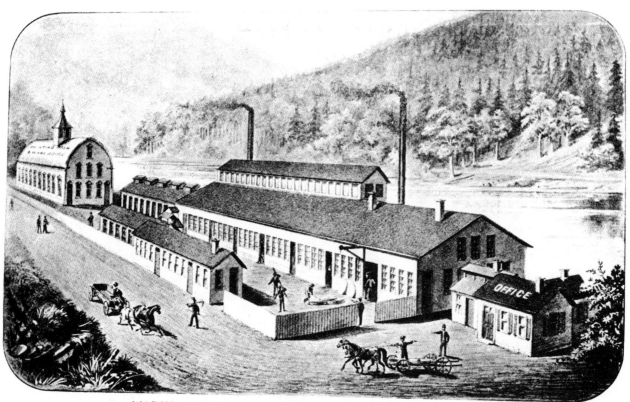

WORKS OF ROBERT MANN & SONS. MILL HALL, PA.

(From an 1888 catalogue, courtesy of Mann Edge Tool Company)

Wheeling, (now West) Virginia. 1834

Meigs, E. H. : East Berlin, Connecticut. Patent (hand axe) #80,868. 1868

Mellinger, Henry : Edge-tool maker, Washington Boro, Lancaster County, Pennsylvania. 1875

Merriman, T. : Waterloo, Wisconsin. Patent (chopping axe) #42,383. 1864

Millen, Joshua : Westborough, Massachusetts. 1832

Miller, George : Cast Steel Edge-Tool Factory, Baltimore, Maryland. 1850

Miller, M. F. : Axe manufacturer, New York City. 1907

Miller, S. & E. : Axe manufacturer, Ticonderoga, New York. 1850

Mills, E. : Edge-tool manufacturer, Philadelphia. 1876

Moorehouse, W. : Buffalo, New York. Patent (axe helve) #47,214. 1865

Morris Axe & Tool Company : Lysander, New York. 1870

Morris, H. D. : Baldwinsville, New York. Patent (manufacturing axes) #98,789; patent (machinery for making axes). 1879.

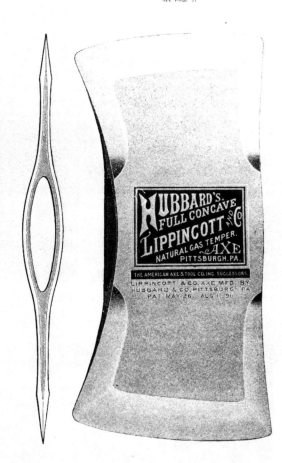

This page from a 1907 American Axe & Tool Company catalogue testifies to the high quality of the axe industry advertising of that day. (Courtesy of Mann Edge Tool Company)

Four labels used by the Mann Edge Tool Company. Each label denoted a different quality axe. (Courtesy of Mann Edge Tool Company)

"UNDERHILL" ICE AXES

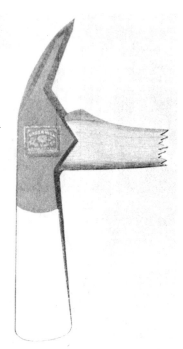

NEW YORK PATTERN
ONE SIZE ONLY:—2½ INCH CUT.
LIST:—PER DOZ. $20.00
WEIGHT:—3 LBS., NOT INCLUDING WEIGHT OF 34-INCH HANDLE.
PACKED:—1 DOZ. IN A CLOSED BOX.
SHIPPING WEIGHT:—71 LBS. PER DOZ.
TO ORDER BY WIRE:—USE CODE, PAGES 5-8.

CHICAGO PATTERN
ONE SIZE ONLY:—2½ INCH CUT.
LIST:—PER DOZ. $28.00
WEIGHT:—3 LBS., NOT INCLUDING WEIGHT OF 34-INCH HANDLE.
PACKED:—1 DOZ. IN A CLOSED BOX.
SHIPPING WEIGHT:—71 LBS. PER DOZ.
TO ORDER BY WIRE:—USE CODE, PAGES 5-8.

"NEW YORK" PATTERN   "CHICAGO" PATTERN
BAYONET POINT

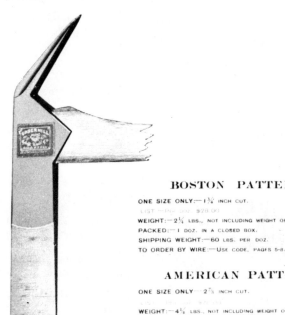

BOSTON PATTERN
ONE SIZE ONLY:—1¾ INCH CUT.
LIST:—PER DOZ. $28.00
WEIGHT:—2¼ LBS., NOT INCLUDING WEIGHT OF 32-INCH HANDLE.
PACKED:—1 DOZ. IN A CLOSED BOX.
SHIPPING WEIGHT:—60 LBS. PER DOZ.
TO ORDER BY WIRE:—USE CODE, PAGES 5-8.

AMERICAN PATTERN
ONE SIZE ONLY:—2⅞ INCH CUT.
WEIGHT:—4¼ LBS., NOT INCLUDING WEIGHT OF 28-INCH HANDLE.
PACKED:—1 DOZ. IN A CLOSED BOX.
SHIPPING WEIGHT:—80 LBS. PER DOZ.
TO ORDER BY WIRE:—USE CODE, PAGES 5-8.

"BOSTON" PATTERN   "AMERICAN" PATTERN

A page from a turn-of-the-century catalogue of the American Axe and Tool Company. (Courtesy of Mann Edge Tool Company)

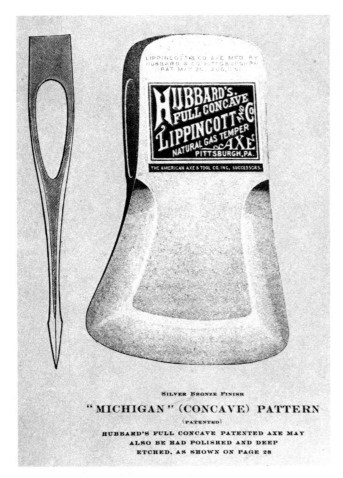

"MICHIGAN" (CONCAVE) PATTERN

An example of a rather rare beveled and hollow-ground felling axe made by the American Axe and Tool Company. (Courtesy of Mann Edge Tool Company)

Morse, John H. : Scythe and axe maker, Shelburne, Massachusetts. 1824

Moulton, Josh. : Blacksmith, Scarborough, Maine. 1832

Munzer, ———— : Patent (axe) #1,330,213. 1920

Napanock Axes and Iron : Wawarsing, New York. 1870

Neal, Elijah : Blacksmith, York County, Maine.

Newmyer & Groff : Pittsburgh. ["Warehouse No. 22 Wood Street, Pittsburgh, Manufacturer of Axes, Railroad Tools, Cast-steel and Steel Point Plated Shovels, Spades and Planters' Hoes, Mattocks, Wedges, Sledges, Etc."]

Newmyer, Groff & Co. : Axes and shovels, Pittsburgh. 1860

Norris, P. W. : Detroit, Michigan. Patent (ox and gruffing axe) #99,784. 1879

North Wayne Tool Co. : Oakland, Maine. 1835–1968

Ogden, M. C. : "Best Cast Steel Axes." 1875

Olds, L. : Maker of axes, hoes, scythes and forks, Oneonta, New York. 1833

Oreley, J. : Ballston Spa, New York. Patent (axe-making process) #9,000. 1852

Osborn & Swan : Edge-tool manufacturers, New York City.

Packard, A. : Axe-maker, Bainbridge, New York. 1850

Palm, ———— : Patent (folding axe) #505,485. 1924

Palm, John A. : Bowmansville, Pennsylvania. 1875

Palmer & Hubbard : Pittsburgh. Patent (manufacturing axes) #188,264. 1871

Palton, Thomas : Edge tools, Philadelphia. 1864

Parker, Samuel : Axe manufacturer, Eaton, New York. 1850

Peabody, W. : Orono, Maine. Patent #118,266. 1826

Peavy Mfg. Co. : Edge-tool manufacturers, Brewster and Oakland, Maine.

Peckman, W. C. : Troy, Ohio. Patent (stone axe) #132,540. 1871

Perkins, Wm. : Blacksmith, Philadelphia. 1789

Peterson, Johnson & Co. : Lockport, New York. 1871

Plumb, Fayette & Co. : Philadelphia. 1969

Poe, George : Mill irons and edge tools. 1825
[From *Maryland Herald & Elizabethtown Weekly Advertiser*, October 14, 1825: The subscriber, living at Sharer's Mill below Funkstown respectfully informs millers of Washington and adjoining counties, that he is prepared to make and turn, mill irons of every description. He will also make edge tools, according to order and on reasonable terms.]

Porter, A. W. : Saint Johnsville, New York. Patent (metal caps for axe helves) #20,192. 1869

Powell, Albert : Axe-maker, Cleveland, Ohio. 1863

Powell & Co. : Cleveland, Ohio. 1845

Powell, F. F. : Patent (axe) #647,609. 1900

Powell Tool and Plaster Company : Cleveland, Ohio. 1877

Powell Tool Company : Cleveland, Ohio. 1872

Pratt, J., Jr. : Charlemont, Massachusetts. Patent (for axe-making machine) 1833

Pratt, Josiah : Claremont, Massachusetts.

Pugh Brothers : Philadelphia. 1868

Quast, E. : Fredonia, Missouri. Patent #96,937. 1869

Rains, R. S. : Patent (axe) #1,282,001.

Rayburn, _____ : Patent (axe) #1,504,-644.

Reeves, _____ : Patent (brush axe) #1,030,429. 1912

Regon, _____ : Patent (axe) #684,908.

Rexford, _____ : Patent (axe box or case) #14,935.

Rexford, O. S. : East Highgate, Vermont. 1812 [*The writer has never seen an axe marked Rexford.* H.J.K.]
[From the Rexford catalogue: *My hand made, ebony half wedge axe. Those in need of a thoroughly first class chopping axe of this shape will find it fully up to the highest standard of excellence, in regard to temper and thorough workmanship. The axe is black and left just as it comes from the tempering forge, showing its various colors of temper essential for the highest grade of cutting qualities, strength and durability. These hand made axes are in great demand by all who follow lumbering and wood dripping.*]

Reynolds, G. : Manchester, New Hampshire. Patent (machine for making axe polls) #320,057. 1878

Reynolds, R. C. : Manchester, New Hampshire. Patent (manufacture of axes) #49,156. 1865

Rockwell, J. N. : Napanock, New York. Patent (hardening axes) #17,639. 1857

Rodgers, Ethan : Patent (design) #21,-253. 1891

Rolf, James : Scarborough, Maine. 1832

Romur Bros. : Edge-tool makers, Gowanda, New York. 1907

Root, E. K. : Collinsville, New York. Patent (punching eyes of axes) #1,027, 1838; patent (axe-dressing machinery) #5,731. 1848.

Rosenthal, Herman : Edge tools, Philadelphia.

Russel, John : Axe manufacturer, Napanock, New York. 1864

Sandridge, _____ : Patent (splitting axe) #1,272,538.

Schanfele, G. : Edge tools, Cleveland, Ohio. 1868

Directory of American Axe Makers

American Axes

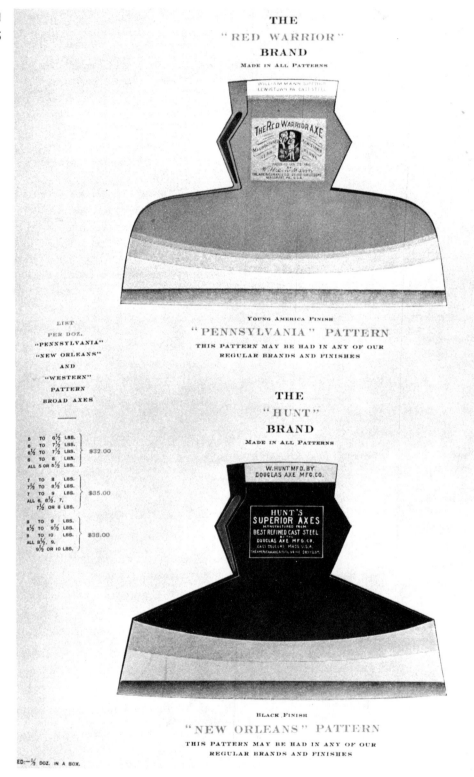

136 Two brands of broadaxes sold by the American Axe and Tool Company around 1910. (Courtesy of Mann Edge Tool Company)

Schwartz, _____ : Patent (axe) #604,693. 1898

Scott, Brother & Co. : Axe manufacturers, Ironton, Ohio. 1853

Sears, J. M. : Vandalia, Illinois. Patent (axe handle) #115,532. 1871

Seeley, _____ : Patent (folding axe) #1,189,005. 1916

Seeley, Nathaniel : Axe manufacturer, Seneca County, New York. 1850

Sellers, Wm. & Co. : Philadelphia. 1876

Selson, Cook, & Co. : Edge tools, Philadelphia. 1868

Selson, George : Edge tools, Philadelphia. 1876

Sener, G. : Axe manufacturer, Lancaster, Pennsylvania. 1841
[Notice in *The Lancaster Union* of November 2, 1841: *G. Sener respectfully informs friends and the public in general, that he has not removed to the West, as has been reported in different sections of the country, by some person or persons, unknown to him, but that he still continues the manufacturing of Edge Tools in all its various branches, at his old established stand, in North Prince Street, Lancaster, where he always keeps on hand an assortment of tools manufactured from the best steel, and which he will warrant to keep a durable edge, equal if not superior to any manufactured elsewhere. Thankful for past favors, he hopes by strict attention to business, and by a careful selection of best materials, to merit and receive a share of public patronage.*]

Shande, _____ : Patent (axe with sledge and detachable cutters) #3,245,094. 1966

Shaw & Burke : Canton, Connecticut. 1831

Shaw, E. : Canton, Connecticut. Patent (for forging axes). 1833

Shertz, John : Edge tools, Lancaster, Pennsylvania. 1873

Shirk, Joseph : Spring Grove, Pennsylvania. 1875

Shotwell, J. : Patent (for sharpening axes). 1799

Silkman, J. H. : Milwaukee, Wisconsin. Patent (axe for wood-splitting machine) #52,216. 1866

Simmons, Daniel : Berne, New York. 1826
[*Daniel Simmons, who had been a blacksmith in the vicinity of Albany, New York, established an axe factory in Berne, New York in 1826. Due to a combination of inadequate financial backing, insufficient water power, and poor transportation to Albany, the growth of the factory in Berne was limited. Ac-*

Directory of American Axe Makers

G. Sener's "eagle" benchmark. (Vincent Nolt collection)

"Underhill" brand axes catalogued by the American Axe and Tool Company in 1907. (Courtesy of Mann Edge Tool Company)

cordingly, Simmons moved his axe establishment to Cohoes, New York, in 1834, which proved an immediate success. The transportation facilities to New York City and the western part of the state must have contributed to the resulting growth of his industry there. Jonas Simmons, thought to have been a brother of Daniel, opened a sales agency of axes at 21 Maiden Lane, New York City about 1842. Four months before his death, Daniel Simmons, with Jonas Simmons, organized the Lockport Edge Tool Co. on October 1, 1860, at Lockport, New York. A newspaper account in 1861 stated that C. A. Olmstead was the agent, and that the factory had 6 forges, 2 triphammers, and stone and emery machinery for grinding and polishing. H.J.K.]

Simmons, J. : Cohoes, New York. Patent (axe-making machine) #9,691. 1853

Simmons & Johnson : Lockport, New York. 1870

Simonds, G. W. : Lynfield, Massachusetts. Patent (mode of securing axe handles) #56,281. 1866

Smith, P. R. : Patent (axe) #246,566. 1866

Smith, R. : Canton, Connecticut. 1832

Smith, Royal : Edge-tool maker, Boston, Massachusetts. 1866

Spafford, Amos : Edge-tool maker, Buxton, Maine. 1815

Spencer, I. M. : Patent (axe) #463,002. 1891

Spiller Axe & Tool Company : Axe manufacturers, Oakland, Maine. 1928

Steiger, J. G. : Cleveland, Ohio. 1868

Stephens, W. M. : Patent (axe) #1,265,276. 1918

Stevens, R. D. : Patent (ox bit) #1,292,531. 1919

Stewart, J. : Money Creek, Minnesota. Patent (mode of attaching axes to handles) #70,284, 1867; patent (fastening handles to axes) #81,308. 1868.

Stohler, J. B. : Blacksmith, Schaefferstown, Pennsylvania. c. 1900

Stone, D. W. : Homer, New York. 1870

Stone, F. D. : Edge-tool maker, Cleveland, Ohio. 1861

Sublett, ____ : Patent (axe and brush hook) #299,876. 1884

Taft, D. : Axe and scythe maker, Windsor, Vermont.

Tatem Manufacturing Co. : Axe-makers, Eastford, Connecticut.

Taylor, Chas. M. : Edge-tool maker, Cheyney, Pennsylvania. 1870

Ten Eyck & Co. : Axe manufacturer, Cohoes, New York. 1863

Ten Eyck, W. J. & Co. : New York City. 1866

Terrill, D. D., Saw Co. : Bangor, Maine.

Tongue, Samuel : Edge-tool maker, Philadelphia. 1876

Towle, Levi : Blacksmith, Westbrook, Maine. 1809

Turner, I. W. : Baltimore, Maryland. Patent (axe-making machine) #2,841. 1842

Turton, Thomas and Sons : New York City. 1866

Tyndale, T. H. : Bellville, Illinois. Patent (attaching handles to axes) #90,412. 1869

Underhill Brothers : Boston. 1866

Underhill Edge Tool Co. : Boston. 1866

Underhill Edge Tool Company : Nashua, New Hampshire. 1907

Updergroff, Abner : Pittsburgh. 1837

Van Slett, ____ : Patent (axe construction) #1,496,250. 1924

Directory of American Axe Makers

American Axes

Vetter, Francis : New York City. 1866

Vetter, Joseph : Edge-tool maker, New York City. 1866

Verrie, J. P. : Philadelphia. 1876

Walden, Hiram : Axe-maker, Wright, New York. 1835

Walker, John : Blacksmith, Scarborough, Maine. 1832

Wall, Thomas : Cleveland, Ohio. 1845

Wallace, ———— : Patent (axe) #1,198,-089. 1916

Warner, H. W. : Patent (axes) #181,-227. 1876

Warner, Hunt & Co. : Douglas, Massachusetts. 1832

[Warnock, Joseph : Galt, Ontario. 1876]

Warren, ———— : Blacksmith, Scarborough, Maine.

Warren Axe and Tool Company : Axe manufacturers.

Warren & Cousins : Edge-tool makers, Buxton, Maine. 1828

Waters, ———— : Blacksmith, Pittsburgh. 1826
[From the *Business Directory of Pittsburgh, 1826*, under "Blacksmiths": *Within the limits of the Corporation (Pittsburgh) there are 24 blacksmith shops, which employ 115 persons. Among these we must notice the establishment of Mr. Waters, near Heron's steam mill, where there are made, weekly, 30 dozen of shovels, 6 dozen axes— making per annum the enormous number of 18,720 shovels and 3,744 axes. Twenty-seven hands are employed.*]

Waters, Oren : Edge-tool maker, Chartier's Creek near Pittsburgh. 1837

Watkins, Evan : Baltimore, Maryland. 1857

Watts, Wm. : Edge-tool maker, New York City. 1866

Weed, Becker & Co. : Axe manufacturer, Cohoes, New York. 1860
[*In 1860 the firm of Weed & Becker was established at Cohoes, New York. William H. Weed and Storm A. Becker had previously been admitted into partnership in association with D. Simmons and Co. Subsequently, this organization again modified its name to Week & Becker Manufacturing Company. A serious fire in the polishing, finishing, and handle shop on May 28, 1875, destroyed their building at a loss of $75,000.00. In 1876 a four-story brick building 140 x 50 feet was completed. The production at that time was 100 dozen axes and 75 edge tools daily, resulting in an annual value of between $200,000 and $400,000. The firm failed in the early 1880's.* H.J.K.]

Weed, Becker & Co. : Edge-tool manufacturer, New York City. 1866

Weed & Cooper : Lockport, New York. 1846

Weed & Hilmer : Edge-tool makers, Lockport, New York. 1846

Weed, N. : New York City. 1866

Wheeler, William : Axe manufacturer, Seneca Falls, New York. 1842

Whipple, M. D. : Douglas, Massachusetts. 1833

White, H. & Sons : Axe manufacturer, Chagrin Falls, Ohio. 1863

White, J. & Son : Scranton, Pennsylvania. 1870

White, L. & I. : Axe manufacturer, Buffalo, New York. 1837

White & Olmstead : Cohoes, New York. 1843
[*In 1843 Miles White joined A. Olmstead of Cohoes, New York, and formed a second axe factory there as the firm of White and Olmstead. White, who was the senior partner, had been traveling agent*

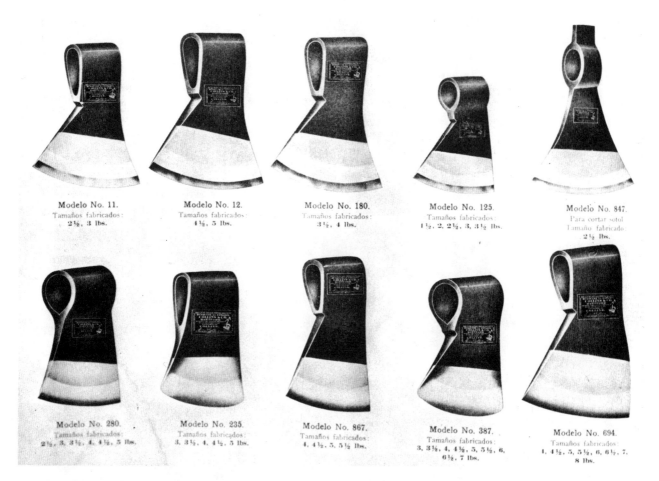

Unusually early forms still in production in the twentieth century for distribution to South America are shown in this Collins Company catalogue of 1919. (Courtesy of Mann Edge Tool Company)

for Daniel Simmons, and had apparently also gained considerable manufacturing experience from his previous employment. More important, he had established acquaintances and contracts with dealers and hardware establishments. The annual value of the products of White and Olmstead as reported in the Census of 1850 were: Axes $72,000 and Edge Tools $18,000. H.J.K.]

White, Thomas : Axe and hoe maker, Ashfield, Massachusetts. 1832

Whitley, J. H. : Patent (axe "or similar tool") #299,876. 1884

Wideman, Henry : Saugerties, New York. 1850

American Axes

Wilcox, C. A. : Patent (axe) #884,250. 1908

Willard, S. : Baltimore, Maryland. 1831

Williams, Filker : Philadelphia. 1866

Williams, G. W. : Maker of "Cast Steel Axes," Cincinnati, Ohio. 1841

Winer, G. W. : Lockport, New York. 1859

Wing, ____ : Patent (folding axe) #738,869. 1903

Winship, ____ : Patent (folding axe) #1,074,764. 1913

Witherell, J. H. : Axe manufacturer, Oakland, Maine.

Woodell, Joseph : "Importer and Dealer in Foreign and American Hardware," Pittsburgh, Pennsylvania. 1847

Wooding, H. C. : Wallingford, Connecticut. Patent (attaching axe handles) #55,437. 1866

Yeatts, ____ : Patent (axe) #910,763.

Yerkes & Plumb : Philadelphia. 1876

Young, C. L. : Patent (axes) #199,342. 1878

GLOSSARY

American axe

Generally, any style or form of axe developed by American craftsmen. Specifically, the American felling or chopping axe developed about the middle of the eighteenth century for cutting down large stands of timber. Its unique quality was a heavy poll, which balanced the weight of the bit and made the felling axe more efficient. The earlier trade axe used in America had no poll. (*see also* Axe types, Broad)

American steel

In the nineteenth century English steel was regarded as superior to American steel because the English technology in steelmaking was more advanced. Yet American patriotic pride was such that if native steel was used in fabricating a tool, it was often so indicated in newspaper advertisements.

Axe

A rectangular metal tool made of iron and steel in the eighteenth and early nineteenth centuries, but today made completely of steel. An eye is formed off-center (except in double-bitted axes) into which a handle is fitted; a wedge keeps the axe on the handle. Used principally for chopping wood; however, specialized forms have been made for carpenters, firemen, ice-cutters, etc. (*see also* Axe types)

Axe blank

A slab of metal from which an axe will eventually be made.

Axe pattern

A plan or template used to draw the desired design on a slab of iron or steel. The term is also applied to the blank in the early stages of development.

Axe types

American: A felling axe.

Biscayan: A trade axe brought to America by the French and made of iron from Northern Spain.

Broadaxe: A hewing axe (such as for shaping timbers) with a long cutting edge ground on one side like a chisel and a wide, relatively heavy bit. All broadaxes have short handles, which are usually canted to one side.

Bull (also *Ox*): An axe with a stud on the poll designed for felling animals in a slaughterhouse. This type of axe is also used by ship and framing carpenters to drive wooden pins or pegs in joining timbers.

Chopping (*see* Felling axe *and* American axe)

Double-bitted: An axe with two cutting edges instead of one edge and a poll. It is used principally by lumbermen or loggers. One edge is usually blunter than the other and is used on frozen woods (which it cleaves or splits) and other work that would blunt a keener edge. This makes it possible to work much longer between sharpenings.

Felling: Designed for felling trees. The American felling axe is a versatile tool good for felling timber and chopping it into lengths once it is on the ground. There are many variations, some better for specific purposes than others. (*see also* American axe)

French trade (*see* Trade axe)

Goosewing: A special type of hewing axe with a canted handle and a form like a goose's wing. Made and used

American Axes

particularly in Pennsylvania in the eighteenth and nineteenth centuries.

Hand: A small axe with a short handle designed to be used with one hand. Often custom-made by blacksmiths for specific woodworking jobs.

Holzaxe (*see* Splitting axe)

Hudson Bay (*see* Trade axe)

Mortising: A long, narrow, chisel-like axe with a blade at a right angle to the haft. Designed to cut square-cornered cavities in timbers (or smaller stock) to accommodate tenons, etc.

Trade: The earliest type of axe used in America, it is distinguished by its lack of a poll. It was brought by the French and Spanish (Biscayan axe) and by the English (Hudson Bay axe) to be traded to the Indians, who probably used it as a tomahawk and general utility axe.

Utility: An all-purpose chopping axe, particularly one such as is available in hardware stores today.

Yankee: An American-style felling axe of the nineteenth and twentieth centuries.

Single-beveled broadaxe. (Dale Pogatchnik collection)

Spanish: A trade-type axe made by the Collins Company in the twentieth century for sale in South America. Also an early trade axe brought to America by the Spanish.

Splitting: Any axe used for splitting timber, usually as a wedge. Splitting axes are usually old felling or even hewing axes that have become too blunt for their original purpose. Some, like the Holzaxe, are designed with extra-heavy, wedge-shaped polls specially for splitting.

Basil

A synonym for the verb *bevel*.

Bevel

A slanting surface ground on the cutting edge of a tool. Tools with one bevel are called single-bevel tools, those having two bevels, double-beveled tools.

Biscayan axe (*see* Axe types)

Bit

The cutting edge of an axe, the part farthest from the handle (except on double-bitted axes, which have the handle in the middle).

Blast furnace

In the eighteenth and nineteenth centuries, a tall rectangular or round building of stone or brick with a hollow cylindrical center into which limestone, charcoal, and iron ore were dumped. The heat generated by a draft from a large bellows caused the iron to be separated from the unwanted substances, both of which were periodically removed before the procedure was repeated.

Blister steel

Steel formed by exposing iron to carbon-bearing substances while hot, the carbon additions being in the shape of blisters on the surface of the metal.

Bloom

A small quantity of iron refined in a bloomery or forge; not entirely pure, but adequate for making tools.

Bloomery

A small hearth into which high-quality iron ore was placed for heating, refining, and subsequent forging into usable wrought iron.

Broadaxe (*see* Axe types)

Bull axe (*see* Axe types)

Cant

The degree of deviation between the axis of the handle and the main axis of the axe.

Cast iron

Usually raw unrefined iron as it is taken from a blast furnace or foundry. Its high carbon content makes it brittle and useful only for objects not subject to impact.

Cast steel

Steel made from blister steel, but melted and cast to effect a uniform distribution of carbon throughout the metal.

Charcoal iron

Iron separated from the ore by using charcoal for fuel. Such iron was excellent for both forging and welding. Except for the cutting edge, all axes of the eighteenth and early nineteenth centuries were made of charcoal iron.

Glossary

The canted handle of a hewing axe. (Vincent Nolt collection)

Double-bitted axe (*see* Axe types)

American Axes

Drop forge

A massive modern machine used to shape axes and other tools.

Dutchman

Generally speaking, a Dutchman is an addition to compensate for an error or inadequate material.

Felling axe (*see* Axe types)

Flashing

A thin metal edge formed where two dies meet in the making of a tool on a drop forge. Usually sheared or ground away with a small ridge remaining.

French trade axe (*see* Axe types)

Goosewing axe (*see* Axe types)

Haft

The handle of an axe or hatchet.

Hand axe (*see* Axe types)

Hardy

An anvil accessory shaped like a chisel and used to bend or cut metal.

Hatchet

An axe to be used in one hand; it can be of any heft but is of a distinctive "hatchet shape." The neck is narrowed under the helve (handle) and eye, and the bit is flared. The edge of the bit is usually straight when new. The common lathing, or half-hatchet, shape is flared only on the back side of the bit (toward the handle), which makes the poll more useful as a hammer. The full-hatchet shape, sometimes used in two-handed chopping and hewing axes, is flared both front and back.

Helve (*see* Haft)

High Carbon Steel

Steel with 1 to 2 percent carbon. A temper can be drawn in such metal to create a hard, tough cutting edge.

Holzaxe (*see* Axe types)

Horse or Pony

A long pole hinged to the floor on which a grinder sat to bring pressure on an axe while it was being ground.

Hudson Bay axe (*see* Axe types)

Mild Steel

Steel with less than 1 percent carbon content. A temper cannot be drawn in mild steel.

Mortising axe (*see* Axe types)

Overcoat

A piece of high-carbon steel slipped over the bit of an axe and welded in place to provide a hard, sharp, long-lasting cutting edge.

Poll

The poll end of an axe is the short, squared portion on the end of the axe opposite the bit. Usually flat but not intended for use as a wedge. Its major function is to properly balance the axe so that the weight is not all in the bit.

Russia iron

A high-quality iron imported from Russia in the eighteenth and early nineteenth centuries.

Spanish axe (*see* Axe types)

Splitting axe (*see* Axe types)

Stud

A metal part or fitting on a tool, such as a hammer or an adze, to drive spikes. Also the part of an axe for felling animals in a slaughterhouse.

Swage

A swage is a shaped steel tool, designed so that, when softer (usually hot) metal is pounded into it, a desired form is produced.

Glossary

Swede's steel

A high-quality steel imported from Sweden in the eighteenth and nineteenth centuries.

Tempering

The procedure of hardening a portion of a cutting tool and then slowly reheating it until only the edge is hard, while the remainder of the tool stays tough and is not brittle. This is a very critical procedure in the making of fine axes.

Trade axe (*see* Axe types)

Utility axe (*see* Axe types)

Wrought iron

Wrought iron is a type of metal from which most of the carbon has been removed. It is highly suitable for forging and welding.

Yankee axe (*see* Axe types)

BIBLIOGRAPHY

American Axe and Tool Company, *Catalogue*. Glassport, Pa.: 1907.

Bishop, J. Leander, *A History of American Manufactures from 1618 to 1860*. Philadelphia: Edward Young & Co., 1861.

Cotter, John L., and Hudson, Paul J., *New Discoveries at Jamestown*. Washington, D. C.: National Park Service, 1956.

Coxe, T., *A View of the United States of America*. Philadelphia: William Ball, Wrigley, and Berriman, 1794.

Diderot, Dennis, et al., *Recueil de planches, sur les sciences, les arts liberaux, et les arts méchaniques, avec leur explication*. Paris: Briasson et al., 1762-1777. 12 vols. These volumes of engraved plates are part of Diderot's *Encyclopédie, ou Dictionnaire raisonné des sciences, des arts, et des métiers par un société de gens de lettres*. Paris: Briasson et al., 1751-1765.

Douglas Axe Manufacturing Company, The, *Price List and Illustrated Patterns*. Boston: Printed by Alfred Mudge and Sons, 1863.

Ellis and Evans, *History of Lancaster County*.

Farmers' Bulletin. Washington, D. C.: U. S. Department of Agriculture, 1962.

Goodman, W. L., *The History of Woodworking Tools*. New York: David McKay Co., 1964.

Great Industries of the United States, The. Hartford, Chicago and Cincinnati: J. B. Burr and Hyde, 1872.

Kelly, William, *A True History of the So-Called Bessemer Process*. Greensburg, Pa.: Boucher, 1924.

Kniffen, Fred, "On Corner Timbering," *Pioneer America*, Vol. 1 (January, 1969), pp. 1-9.

McLane, Louis, ed., *Documents Relative to Manufactures in the United States and Transmitted to the House of Representatives by the Secretary of the Treasury*. Washington, D. C.: U. S. Treasury Department, 1833. 2 vols. (U. S. 22d Congress, 1st Session, 1831-32, House Document #308)

McLaren, Peter, *Axe Manual*. Philadelphia: Fayette R. Plumb, Inc., 1929.

Mann Edge Tool Company, *Axes*. Lewistown, Pa.: 1897.

Mercer, Henry C., *Ancient Carpenters' Tools*. Doylestown, Pa.: The Bucks County Historical Society, 1951.

Moxon, Joseph, *Mechanick Exercises: or Doctrine of Handy-Works*. London: Printed for Dan Midwinter and Thos. Leigh, at the Rose and Crown in St. Paul's Church Yard, 1703.

Overman, Fredrick, *The Manufacture of Steel*. Philadelphia: Moss and Brother, 1854.

Rayl & Company, *Illustrated Catalogue and Price List of Tools*. Detroit.

Russel, Carl P., *Firearms, Traps, and Tools of the Mountain Men*. New York: Alfred A. Knopf, 1967.

Scientific American, Vol. 1 (October 29, 1859), pp. 284-5.

Shoemaker, Alfred, ed., *The Pennsylvania Barn*. Kutztown, Pa.: Pennsylvania Folklife Society, 1959.

Tryckare, Tre, et al., *The Viking*. Gothenburg, Sweden: Cagner & Co., 1966.

INDEX

American Axe & Tool Co., 41, 116; ills. 131, 133, 136, 138
American Fork & Hoe Co., 116; ill. 75
Axe(s):
  abuses, 103; ill. 102
  advertising, 43, 48, 101, 102, 104-105; ills. 127, 131
  -bars, 35, 45
  blank, 143
  brands. (See Climax, Diamond, Jack Frost, Lewis, Lippincott, Perfect, Pine Tree, Pioneer, Red Warrior, Superior, Underhill, Victory)
  care of, 101-106; ills. 102, 105. (See also Filing, Grinding, Guarantees, Handle fitting, Honing, Warranties)
  collectors. (See Collectors)
  manufactories:
    18th Century, 16-29;
    19th Century, 30-51;
    20th Century, 52-56.
    (See also Directory of American Axe-Makers, 115-142)
  manufacturing methods:
    early axes, 5-15;
    18th Century, 16-29;
    19th Century, 30-51;
    20th Century, 52-56.
    (See also Blacking, Edging, Grading, Grinding, Handle fitting, Overcoating, Packaging, Polishing, Tempering, Testing)
  parts. (See Bit, Eye, Handle, Poll)
  patterns, specific. (See Black Raven, Boston, Chicago, Jersey, Kentucky, Ohio, Maine, Michigan, Muley, New England, New Orleans, New York, Pennsylvania, Pittsburgh, Ship Carpenter, Spanish, Western, Wisconsin, Yankee) (See also Patterns)
  types, 143. (See also Biscayan, Broadaxe, Bull, Carpenter, Double-bitted, French trade, Felling, Goosewing, Holzaxe, Ice, Mast-makers, Mortising, Non-woodworking, Posthole, Shop, Single-bitted, Spanish, Turpentine, Utility)

use of, 107-114; ills. 108, 110, 112, 113. (See also Bucking, Dovetailing, Corner-timbering, Felling, Hewing, Mortising, Notching, Squaring)

Basil, 26, 47, 144
Bayeux Tapestry, axes in, 9
Beatty, John C., 116; ill. 73, 83
Bellows, 33
Benjamin, Timothy, 116; ill. 88
Bevel, 106, 144. (See also Double-Bevel and Single-bevel)
Bick iron, 28
Biscayan axe, 11, 13, 143. (See also French trade axe)
Bit, 7, 9, 11, 18, 20, 35, 38, 48, 144-145; ill. 61. (See also Double-bitted and Single-bitted)
Bitumen, use of, 6
"Black Diamond" finish, 39
Blacking, 32
"Black Raven" pattern, ill. 75
Blacksmith(s), role of, 3, 9, 16, 18, 19, 20, 22, 45
Blank, 23
Blast furnace, 17-18, 145
Blister steel, 45, 145
"Bloom," 17, 145
Bloomery, 17, 18, 145
Bolts, 109
Borax, 36
"Boston" pattern, ill. 133
Bow-drill, 6
"Box," 108
Brady, William, 117; ill. 85
Brands, specific. (See Axe brands)
Brittleness, problem of, 36
Broadaxe(s), 46, 47, 48, 143; ills. 65, 66, 67, 68, 69, 70, 71, 72, 88, 117, 120, 136, 144. (See also Hewing axe)
Bronze, use of in axes, 7, 8, 9; ills. 7, 8
"Bucking," 109
Bull axe, 143; ill. 97
Burr, 106

Cant, 26, 145; ill. 145
Care. (See Axes, care of)
Carpenter's axe, ill. 15

Cast steel, use of in axes, 29, 39, 45, 145; ills. 44, 71, 72, 78, 81, 132
"Celts," 6
Charcoal, 32, 33
Charcoal iron, 18, 45, 145
"Chicago" pattern, ill. 133
Chinking, 113; ill. 113
"Climax" brand, ill. 132
Coal, 30, 33
Collectors, 4, 47, 115. (See also Portfolio)
Collins Company, 30, 32, 34, 38, 41, 42, 118; ills. 32, 38, 53, 72, 119, 141
Collins, D. C., 30, 31
Collins, Samuel W., 30, 36; ill. 31
Cooper's axe, ill. 83
Copper, use of in tools, 7
Corner-timbering, 111, 113, 114; ills. 112, 113

Decarbonization, 20, 36
"Diamond" brand, 46
Double-bevel, 144; ills. 66, 83, 138. (See also Bevel)
Double-bitted axe, 9, 47, 145; ills. 46, 56, 82, 131
Douglas Axe Company, 23, 43, 45, 119; ills. 78, 80
Dovetailing, 113, 114; ill. 113
Drop forge, 52, 55, 146
Dunlop and Madeira Co., 119; ills. 69, 120
Dutch, as distributors of axes, 13
"Dutchman," 146. (See also Throatpiece)

Edge-tools, 30
Edging, 22, 36, 39. (See also Grading; Grinding; Steel, use of; Tempering, Testing)
Emerson and Stevens Co., 102, 121; ill. 46
"English type" axe, ill. 82
"Eoliths," 5, 9
European axes, in America, 2, 5-15; ills. 7, 8, 10, 12, 14, 15
Eyepin, 20, 35, 38, 46, 53, 55

Felling, 34, 107, 108-109; ill. 108
Felling axe, 2, 23, 47, 143; ills. 2, 24, 25, 46, 54, 74, 75, 76,

149

Index

77, 78, 79, 82, 127, 131, 134, 141
Finishing, 38, 39, 55
Filing, 106
Flashing, 55, 60, 146
Flux, 20, 36
Forge, 17, 18, 32, 33, 35, 41; ills. 32, 43
Forging, 32, 33, 34, 35
French, as distributors of axes in America, 11, 13
French trade axe, 11, 13, 143, 144; ills. 14, 59, 60, 61
Fullerton of Boston, 45; ill. 44
Furnace. (See Blast furnace)

Germans, as distributors of axes in America, 26
Goodman, W. L., 1, 9
Goosewing axe, 19, 26, 28-29, 143; ills. 27, 28, 62, 63, 64
Gothic pole-axe, ill. 10
Grading, 48
  by color, 36
Grinding:
  hints on, 105; ill. 105
  manufacturing step, 32, 33, 36, 38, 42, 45, 55; ill. 38
Gristmill, 33
Guarantees, 48, 49, 101-103

Haft. (See Handle)
Halberds, 28
Hammerman, 36
Hand axe, 144; ills. 81, 83, 89. (See also Utility axe)
Handles, 6, 9, 20, 23, 26, 42, 47, 50; ills. 49, 76, 145
  fitting of, 6, 33, 104-105
Hardy, 146
Hatchet, 11, 146; ill. 80
Head. (See Poll)
Heading axe, ill. 10
Helve. (See Handle)
Hewes Tool Company, 123; ill. 43
Hewing:
  axe, 1, 11, 14, 23, 26, 47, 107; ills. 73, 89, 145
  "to the line," 109, 111; ill. 110
Hickory, use of in handles, 50; ill. 49
High-carbon steel, 35, 36, 46, 53, 146
Holmes Axe factory, 101, 123
Holzaxe, 144; ill. 77
Honing, 106
"Horse," 38, 55, 146; ill. 38
Hoster, Joseph, ill. 97
Hubbard axes, 124; ills. 131, 134

Hudson Bay axe, 13, 144
Hunt, Warren, axe, 124; ills. 78, 125, 136

Ice axes, ills. 95, 133
Iron, use of in axes, 2, 5, 9, 11, 13, 16, 17, 18, 20, 22, 25, 26, 29, 33, 34, 35, 42, 45, 46, 52, 53, 146; ills. 66, 73

"Jack Frost" brand, 48
Jay brand, 124; ill. 132
"Jersey" pattern, 23, 47

Kelly Axe Company, 42; ills. 74, 75, 127
"Kentucky" pattern, 23, 30, 31, 33, 47
Knapp, H., ill. 81
Krause, George, ill. 128

Labels, for axes, ill. 132
Lathes, use of, 50-51, 104
"Laying" an edge, 18, 22
Lehigh coal, 30, 33
"Lewis" brand, 48
"Lippincott" brand, 128; ills. 131, 134
Log-dog, 111; ill. 110
"Lumberman's Pride" pattern, ill. 46

Machetes, 33
"Maine" pattern, 47
Maker's mark, use of, 2, 4, 38, 41, 47; ills. 60, 62, 63, 67, 70, 71, 72, 73, 84, 86, 88, 97, 128, 137
Mandrel. (See Eyepin)
Mann Edge Tool Company, 3, 39, 41, 48, 129
Mann, Harris, 41
Mann, Harvey, 41
Mann, James H., Company, 41
Mann, J. Fearon, 41
Mann, Robert & Sons, 53, 129; ills. 129, 130, 132
Mann, William, family, History of the, 39
Mann, William, Jr., 41
Mann, William, Sr., 39
Manufactories. (See Axes, manufactories)
Manufacturing. (See Axes, manufacturing methods)
Mast-maker's axe, ill. 87
Mercer, Henry, 26

"Michigan" pattern, 47; ill. 134
Middle Ages, 9, 11; ill. 66
  axe shapes, ill. 12
Mild steel, 34, 52, 53, 146
Mortise-and-tenon, 114
Mortising, 107
  axe, 144; ills. 84, 85, 86
"Muley" pattern, 47

"New England" pattern, ill. 138
"New Orleans" pattern, 47; ills. 69, 136
"New York" pattern, ill. 133
Non-woodworking axes, ills. 90, 91, 92, 93, 94, 95, 96, 97, 98, 99
Notching, 108, 112, 113; ills. 108, 112

"Ohio" pattern, 47; ill. 138
Open-shop declaration, ill. 40
Overcoating, 53, 55, 146; ill. 54

Packaging, 32, 55; ill. 129
"Palstaves," 8, 9
Patent(s), 30, 115; ill. 125
Patterns, 20, 32, 33, 35, 46, 143; ill. 56. (See also Axe patterns)
Peavy Company, 134; ill. 56
"Pennsylvania" pattern, 47; ill. 136
"Perfect" brand, ill. 74, 127
"Pine Tree" brand, ill. 46
"Pioneer" brand, ill. 46
"Pittsburgh" pattern, 47
"Plating-out," 35
Plumb, Fayette R., Company, 104-105, 135
Polishing, 32, 33, 38, 55
Poll, 9, 20, 23, 34, 35, 39, 42, 146; ills. 2, 22, 24, 25, 54, 59, 77, 79, 80, 86, 94
Poll-less axe, ill. 21

"Red Warrior" brand, 41; ill. 136
Rivets, in axes, 25-26; ill. 93
Russel, Carl P., 13
Russia Iron, 32, 146

Saddle notching, 112; ill. 112
Saugus (Mass.), 18
Scarf, 20, 23. (See also Flashing)
Scoring, ill. 110
Sener, G., 29, 45, 137; ills. 28, 70, 84, 137
Shaping, 34, 36, 46

Index

Sheath, 109
"Ship Carpenter" pattern, ill. 138
Shop axes, 58; ills. 83, 84, 85, 86, 87, 88, 89
Single-bitted axe, 144; ill. 134. (See also Bit and Double-bitted)
Single-bevel, 144; ills. 73, 144. (See also Bevel)
Slug, 35. (See also Throat piece)
South America, use of axes in, ill. 141
Spanish, as distributors of axes in America, 13
"Spanish" axes pattern, 20, 47, 144; ill. 141
"Squaring" a timber, 109-111; ill. 110
Steel:
  American, 143
  production of, 18-19, 29, 48
  Swedes, 147
  use of in axes, 13, 16, 17, 18, 20, 22, 23, 25, 29, 33, 42, 45-46, 48, 53, 101; ills. 28, 44, 54, 61, 66, 68, 73, 80, 98
  (See also Cast, High-carbon and Mild steel)
Stohler, J. B., 139; ill. 86
Stone Age, 5, 13
Stone, use of in axes, 5, 6, 32
Stud, 146; ill. 97. (See also Bull axe)
"Superior" brand, ill. 132
Swage, 146

Tempering, 32, 33, 34, 36, 42, 55, 147
Testing, 43, 48, 102
  machine for, ill. 125
Throat piece, 35, 53
Timber. (See Axes, use of)
Trade axe, 11, 13, 15, 20, 23, 144; ills. 14, 59, 60, 61
Transitional axe, ill. 22
Triphammer, 17, 32, 33, 34, 35, 36, 42, 46, 47, 52
Turf axes, ills. 90, 91, 92, 93, 94
Types. (See Axe types)

"Underhill" brand ice axes, ills. 133, 138
U.S. Patent Office. (See Patents)
Utility axes, 11, 15, 144; ills. 15, 37, 54, 77, 78, 80, 81, 82

"Victory" brand, ill. 46
V-notching, 113

Warranties, 43, 49, 102-103
"Wedge" pattern, ill. 79
Weed, Wm., 140; ill. 71 (top)
Welding, 34, 35, 36, 45, 46
"Western" pattern, 47; ill. 131
"Wisconsin" pattern, 47
Working conditions, 33, 35, 38; ill. 38, 40
Wrought iron, 147

"Yankee" pattern, 30, 31, 47, 144; ill. 131

Henry J. Kauffman was born in York County, Pennsylvania, in 1908; graduated from Millersville State Teachers College (now Millersville University) in 1932; and taught industrial arts in high schools in Connecticut and Delaware County, Pennsylvania.

In 1937, Kauffman earned a M.A. from the Pennsylvania State University. For thirty-one years, he taught metalworking including blacksmithing and silversmithing in the Industrial Arts Department at Millersville University becoming the only teacher without a doctorate to earn a full professorship.

Professor Kauffman's knowledge of antiques, hardware, and guns soon made him the authority on the Pennsylvania German material culture including its architecture. His museum-quality works in copper, brass, iron, pewter, and silver attest to his great skill as an artist and artisan, and can be seen at the Rock Ford Kauffman Museum, Lancaster, Pennsylvania, and Yale University's permanent silver collection.

The author of more than 300 articles in scholarly journals and historical publications on antiques, folk art, and architecture, Kauffman continues authoring books including: *Early American Copper, Tin, and Brass* (1950), *Early American Gunsmiths, 1650 to 1850* (1952, reprinted 1965), *The Pennsylvania Kentucky Rifle* (1960, reprinted 1965), *American Ironware, Wrought and Cast* (1966, reprinted 1978), *The Colonial Silversmith* (1969), *The American Pewterer* (1970), *American Axes* (1972, reprinted 1994), *American Fireplaces, Chimneys and Mantles* (1972), *American Andirons and Other Fireplace Tools* (1974), *The American Farmhouse* (1976), *Pennsylvania Dutch American Folk Art* (1964, reprinted 1993), and *Architecture of the Pennsylvania Dutch Country, 1700-1900* (1992).